FPGA Simulation

FPGA Simulation

A Complete Step-by-Step Guide

Ray Salemi

For my friends in Waltham.

Acknowledgments

From the moment I first thought, "You know, there ought to be a book that teaches FPGA simulation" to the moment I completed it, I was blessed by people who appeared and said, "You're right. How can I help?"

My first thanks go to the members of the online review group are one of the big reasons the book got written. They were my writing audience for a year and a half before the book was completed. This audience exerted a subtle pull over me, forcing me to get up early each day and move closer to releasing yet another chapter. It was that daily pull that eventually drew the book out of me. Thank you all.

Thank you Michelle Lange and Mike Mintz, who read the chapters carefully and pointed out, sometimes in paragraph-by-paragraph detail, where the book could be improved. They were the ones who said, "Yeah. That thing you thought you explained clearly—it's not so clear." The clarity of the book, such as it is, is a tribute to their readership. The confusing parts are my fault.

Thanks to the hard-core technical reviewers, such as Jon McDonald, Robert Jeffery, and James Keithan. They read the chapters and said things like, "That technical word you keep using—I do not think it means what you think it means."[1] or "Have you considered being a little less dogmatic." They helped keep the book on the correct technical path.

Mark Glasser got me over the final hurdle. He was in his car, driving to the airport, when he took the time to call and help me through a problem. Thank you, Mark!

Thanks to those who gave encouragement, such as Bob Beckwith, Lyle Benson, and Shashi Bhutada. They said, "This does just what you think it does," and it was nice to hear that.

Thank you to Harry Foster for his kind and wonderful foreword.

1. With apologies to the film "A Princess Bride."

Martin Keirnicki, Bob Pellegrino, and Von Wolf added their experienced engineering voices. They took the book's ideas and applied them to their jobs. Their feedback let me know that I was on the right track, and that this book would be valuable. Thank you, gentlemen.

Thank you to my smart and lovely wife, Karen Salemi, who spent hours poring through dense chapters of technical matter to fix my grammatical errors. Her sharp eye has saved me from embarrassment many times.

I owe an enormous debt of gratitude to Mike Mintz (again), Henrik Scheuer, and Edward Arthur who dug through the almost-final version and found edits that gave the book its final polish.

An enormous thanks to my technical editor, Michael Meyer, who is, without a doubt, the best editor I have ever met. His passion, dedication, and attention to detail were awe-inspiring. If you're writing a book, make sure you talk to Michael about editing it.

And, of course, thank you to Mentor Graphics for making this book possible. Mentor loaned me the software I used to create the examples, and made it legally possible for me to use my work at Mentor in the creation of this book. In addition, my coworkers were a constant source of encouragement, supporting me through the whole process.

Thank you, all!

Foreword

There's a Latin proverb that states, "Times change and we change with them." This is certainly true within the electronics industry. For example, today we are witnessing a phenomenal increase in FPGA design starts. In comparison, ASIC design starts are declining. Although FPGAs have traditionally been relegated to glue logic, low-volume production, or prototype parts used for analysis, this is no longer the case. Gate count and advanced features found in today's FPGAs have increased dramatically to compete with capabilities traditionally offered by ASICs alone. This change in FPGA capabilities has resulted in the emergence of advanced FPGA system-on-chip (SoC) solutions, which includes the integration of third-party IP, DSPs, and multiple microprocessors—all connected through advanced, high-speed bus protocols. Accompanying these changes has been an increase in design and verification complexity, which traditional FPGA flows are generally not prepared to address. This raises many questions related to traditional FPGA methodologies and the changes required to address this increased design and verification complexity.

When I was young, my dad liked to tell me, "If you don't ask the question, you won't get the answer." Today, I have a deeper understanding of his message—and I have learned to apply his wisdom to many areas in my life. For example, as an engineering manager, I have learned (perhaps the hard way) the importance of first asking the question, "Is verification complete?," before proceeding to my next project milestone. Although on the surface this might seem like a trivial question to ask, experience has revealed that it is not necessarily an easy one to answer. In fact, to answer this question with confidence requires a deep understanding of the processes, skills, infrastructure, and metrics used to measure success within an organization.

In *FPGA Simulation*, Ray Salemi presents a novel seven-step approach to improving FPGA verification methodologies, with the goal of building the "ultimate test bench." What makes Ray's approach so valuable is that readers can adopt his various suggestions incrementally, without having to abandon their existing methodology. His common-sense approach is both informative and fun to read. After completing *FPGA Simulation*, the reader is equipped with the necessary tools to implement a methodology that helps provide an affirmative answer to the question, "Is verification complete?"

Engineers will continue to "ask the questions" within the evolving world of FPGA design, and as in any field of practice, asking the question is critical to finding the answer. Ray has eagerly and competently addressed the important question, "Is verification complete?" And in doing so, he has created a book that will serve as a valuable FPGA simulation reference for years to come.

Harry Foster
Chief Verification Scientist
Mentor Graphics Corporation

Preface

It has dawned on me that I've become an old hand in the electronic design industry. I've been working with digital design for over twenty years, starting with programming Altera EPLDs at Prime Computer, moving on to leading the Verilog-XL customer support team at Gateway, passing through managing an engineering tools team at Sun Microsystems, jumping into being a product marketing manager for NC-Simulator and Incisive at Cadence, and landing happily as an Application Engineer Consultant with Mentor Graphics.

Through all that time there has been a consistent theme that's driven digital design and my career:

Chips are getting bigger.
Chips are getting faster.

This was true when the chips in question had 50,000 gates, and it's true today when the largest chips have more than 20,000,000 gates. When I designed with Altera parts at Prime, each pin had its own logic array, and routing meant that you had made an equation with too many terms and had to steal some terms from the next pin. Back then, I would program my EPLD, test it in the lab, find a bug, erase the FPGA under a UV light, and then program it again.

Ah, those were the days.

Today's FPGAs are nothing like my little EPLD from 1988; they have millions of gates, thousands of product terms, and routing issues that we didn't dream of back then. They are highly complex devices with embedded microprocessors and advanced I/O interfaces.

And yet, I see engineers trying to debug these enormous devices the way I debugged small devices back at Prime. They program them, take them to the lab, find a bug, and try again. This has led to needless suffering for these engineers and their families, as they try to do the nearly impossible—debug a million-gate device with a logic probe.

It doesn't have to be that way.

I wrote this book because I've met many engineers who get stuck in the lab trying to debug a complicated FPGA that they don't know how to simulate. I've wanted to recommend a book on FPGA simulation to help them, but I could never find one. There simply weren't any books written to help FPGA designers learn to simulate their RTL before debugging it in the lab.

The problem is not that there aren't any books on RTL verification. Engineers who design ASICs have been verifying their designs for years—after all, a bug in an ASIC can cost upward of a million dollars. As a result, there are many verification books written for ASIC designers; the problem is that these books are very complicated and they assume years of simulation experience.

There was no simple book that taught simulation from the ground up—until now.

I wrote *FPGA Simulation: A Complete Step-by-Step Guide* for the engineer who has decided to start simulating—you, I hope. This guide assumes that you have little or no experience with FPGA simulation, and no time designated for it in the schedule. Given those conditions, this book provides you with a first step that takes no time at all and moves on from there—much like modern language instruction that dispenses with the complicated grammar and gets you quickly to the point where you can order dinner in a foreign country.

Which brings us to the sequence of the steps on the book's cover. I'm constantly asked, "Would someone really follow those steps in that order on a project?" And the answer is, "No." An engineer who knew all the simulation skills in this book would write a test plan first. Then that engineer would design a test bench and a coverage model, and finally run simulations until all the coverage goals were met.

The steps on the cover are not ordered according to how you'd use them in a single project; instead, they are ordered with respect to *adoption*. They are designed to bring people who do little or no simulation to the point where they can use the latest simulation techniques.

You can add these seven steps to your repertoire by adding one step per project. Also, you can stop adding steps whenever you want, because each level on the staircase is self-contained.

The first step, Code Coverage, is the easiest. You simply turn on a feature that is already built into the simulator. Knowing your code coverage will go a long way toward helping you get a handle on bugs in the lab (hint: look at the uncovered code) and will provide a goal. Once you've taken that first step, you can decide whether you want to take the next one.

As much as I love books, I have to admit that reading a book will not, by itself, teach a new skill such as FPGA simulation. In addition to reading the book, the reader must be able to see examples and ask questions. Without that interaction, it is very difficult to learn a new skill.

That's why I created the website www.fpgasimulation.com as a companion to this book and a place for engineers to help each other learn. The website has a single-minded focus:

The website www.fpgasimulation.com provides detailed technical answers and examples to engineers simulating FPGA.

The website has the following resources to help you learn how to simulate:

- A weekly email newsletter that delivers a technical tip.
- Articles that show engineers how to implement new techniques.
- Example designs, including every example in this book.
- Complete, open-source, test benches for common designs.
- Forums where engineers can get answers to specific questions.

When you start reading this book, go to www.fpgasimulation.com to get the source code for the examples. Then you can simulate them yourself as you read about the techniques. Also, sign up for the forums and ask questions. Join the community of learners.

Here is a rock-solid prediction for the future: FPGAs will be bigger. Soon it will be impossible to meet a schedule simply by debugging an FPGA in the lab. If you begin adopting the steps in this book, you'll have the skills you need to conquer the significantly larger design challenges that await us in the future.

I look forward to seeing you on www.fpgasimulation.com!

Ray Salemi
Framingham, Massachusetts
December 31, 2008

Contents

15

Creating an RTL Test Bench ***181***

16

Creating a Transaction Predictor ***199***

28

Introduction to Covergroups ***329***

29

Defining Data Bins ***339***

1 The Boiled Frog

In This Chapter:

- *A Boiled Story*
- *The "Verification Complete" Milestone*
- *Taking it Step by Step*
- *The Seven Steps to the Ultimate Test Bench*

There's a famous story about how best to boil a frog. As the story goes, if you throw the frog into boiling water it will jump right out, but if you put the frog in cool water and heat it slowly, then the frog will stay there until it dies.

Personally, I think that a real frog thrown into a real boiling pot of water would suffer a horrible, watery death, though I agree with the lesson of the story. We are willing to put up with increasing difficulties as long as they increase slowly. Then one day, we look around and realize that we have been cooked.

As FPGA designers, we are the frogs and we are cooked. Our projects usually have a schedule milestone called "In the Lab" and a three-week task called "Lab Bring-up." Three weeks is laughable. We should replace the task "Lab Bring-up" with a task called "Death March," and "Death March" should have no end date.

Once upon a time, getting into the lab really meant that you could expect to ship your FPGA three weeks later. That was back when we were aggregating board logic onto an FPGA whose most complex logic was an address MUX.

We didn't simulate our designs back then because there was little point. The designs were simple and it would have taken longer to create a test bench than to fix the design in the lab. Besides, being in the lab was more fun.

As time went on, however, Moore's law began to turn up the heat. The FPGAs got larger and our lab bring-up time got longer. We began putting more complex logic into designs, but we continued to neglect simulation. Today, we have huge designs and almost no simulation.

This is obscenely expensive. The most obvious cost is debug time. Three weeks in the lab costs about $12,000 per engineer in fully loaded costs. Three months in the lab costs about $50,000 per engineer. That's an incremental difference of $38,000 per engineer over regular debug time.

Then, of course, there is the risk. If we knew that it would take three months to bring up a fully loaded FPGA, we could just write that into our schedule. It would be expensive, but at least it would be predictable. However, that's not the case. Three months of lab time is an average with a high standard deviation; it could easily be longer. The only thing worse than knowing your project will take three months is not knowing how long it will take at all.

A missed schedule is only one cost of neglecting simulation. A product recall is another, much more concrete cost. We know two things about lab debug:

1. We never see the bug that causes a recall in the lab, because if we had seen it we would have fixed it.
2. We can never know whether we've fully tested our RTL.[1]

These two problems make it very likely that a complex, unsimulated, lab-debugged FPGA will see problems in the field.

Finally, there is the human factor. Debug death marches have high — and completely unmeasurable — costs, in stress, missed family events, lack of weekends, burnout, closed market windows, and short-circuited careers.

In short, we are boiled frogs, and it's time to heave our bloated carcasses out of the pot before its too late.

1. "RTL" in this book means synthesizable Verilog or VHDL code.

1.1 A Boiled Story

The scene is a status meeting. A group of engineers sits in a conference room while their engineering director takes the status over a speaker phone that looks like an alien space ship. The engineering lead sits next to the phone:

“So, is the design in the lab yet?” asks the director.

“No. Not yet,” says the lead.

“The schedule says you’re supposed to be in the lab next week. Will you be there?”

“Absolutely.”

A week passes. We are back in the conference room with the engineering lead sitting next to the phone.

“Are you in the lab yet?” asks the director.

“Yup. We got in yesterday.”

“Excellent! Good work everyone. The schedule says we’ll be done with bring-up in three weeks. Let’s hit the milestone and ship!”

As the director hangs up the ten engineers look at each other around the table. It’s true that they are in the lab. The FPGA is on the board on a lab bench. The milestone is complete. Of course, the FPGA doesn’t work. It might as well be a lump of coal. The death march has begun.

Six weeks pass. The engineers have been working round-the-clock shifts to try to debug the FPGA. They’ve been knocking bugs off at the rate of one per four days, but they are three weeks late.

As a result of these failures, the engineering director has flown in to sit in the room during the status meeting. He wants to be there with his engineers as they give their reports because he wants to look into their eyes.

He fixes the lead with a stare and asks, “Are we out of the lab yet?”

“No,” the lead looks down at his notes as he answers.

"What's the problem?"

"We're having trouble with a bug."

"Is this the same one as last week?"

"No, we fixed that one. There's another one."

There's silence.

"When will you be out of the lab?" asks the engineering director quietly.

The lead doesn't like the quiet tone but he answers truthfully, "I don't know."

The lead and the engineering director look at each other. They had this same conversation last week. They'll have this same conversation next week. Odds are they'll be having this conversation for the next two months.

Shift the scene to an upper-level management meeting. They are discussing the engineering director. One VP looks at the other and says, "I don't know about him. He just doesn't seem able to deliver."

Another promising career has been boiled away by the lab.

1.1.1 Root Cause Analysis

What destroyed this project and this engineering director's career? Was it improper design techniques? Perhaps. Was it the tools? Probably not. Was it lack of dedication? No. Was it a lack of proper verification? Surprisingly, no. While this was the immediate cause, it was not the root cause.

The root cause analysis shows that the engineering director's career was rightfully stalled because the problem was his fault.

His schedule destroyed the project. Not the dates, but the milestones. The engineering director had created a schedule with a milestone that said "FPGA in Lab," followed by a "Lab Bring-up" task. He had neglected a critical step.

The schedule was missing the one milestone that would have ensured success: "Verification Complete."

1.2 The "Verification Complete" Milestone

We need to achieve complete functional verification before we get to the lab, and we need a milestone on the schedule called "Verification Complete" to make that happen. What gets measured gets done. If the Director of Engineering in our story had been asking, "Are you done with Verification?," he'd still have a career.[2]

1.2.1 What do you mean by "Verification Complete?"

Schedule milestones need to be clearly defined. We need to be able to say, for a fact, that the milestone has been completed before we check it off.

Verification has a slippery history in this regard. There are many ways to measure verification. For example, you could measure how many tests you wrote, or how long you ran the simulation. You could even measure whether you've run out of time and are forced to go into the lab. None of these are real measures of complete verification.

There are two questions whose answers tell us whether we've completed verification:

- Have we simulated the entire design?
- Are we still finding bugs?

If we've simulated the entire design and we aren't finding any bugs then the verification process is complete.

Many engineers tell me that the goal of simulating an entire design is unrealistic. Also, it's true that you can't simulate every possible set of inputs to an ALU, and you can't simulate every memory location.

Though you can't simulate everything, you can find the important aspects of your design and simulate those. You can simulate all the addressing modes, or commands, or combinations of the two. You can make sure that your router can pass traffic among all its ports.

2. Lest you think this story is exaggerated, I am on the web right now looking at a job opening to replace a Director of Engineering who was done in by verification woes.

This type of simulation is not new. The chip design industry has been doing successful verification in the ASIC world for twenty years. It's possible. It's being done today. Fortunately, as FPGA designers we don't need to go to the lengths of ASIC designers, but we still need to verify all the important functions in our design before we touch it with a logic analyzer.

1.3 Taking it Step by Step

Some books preach radical revolution. They say that if you don't change everything today, you will be left behind in a morass of your own regret. This is not one of those books.

Engineering is a tricky business. It's difficult to get billions of bits to line up and march in order. Frankly, given the complexity of what we do, it's amazing that technology works at all. Despite the complexity, however, we make our projects work over and over again.

We don't do it by radical innovation. Instead we take something that works and add something to it. This way, step by step, we improve our design skills and deliver more-complex projects.

The same holds true of improving verification techniques. While many techniques are available to us, it would be folly to try to implement them all at once. Instead we should adopt a staircase model. We should move from step to step, incrementally adding capability on each project. That way we'll be able to look back down the staircase and be amazed at how far we've come.

1.3.1 The Quest for the Ultimate Test Bench

> *"If you don't know where you are going, you might wind up someplace else."*
> – Yogi Berra

Before we start climbing the staircase, we should know what's up there. What can we expect at the top of the verification staircase? In the best of all worlds our test bench will do three things:

- Generate its own stimulus so we don't have to write many tests.
- Check the output of the device under test (DUT) to make sure it is correct.
- Tell us when the DUT has been completely tested.

This is the Ultimate Test Bench, and it is within your grasp because you are holding this book. *FPGA Simulation* will take you, step-by-step, from a simple directed test bench environment to the Ultimate Test Bench.

Each step builds upon the previous steps, so that you don't need to be a radical to get there. You can just add a step or two to each project until you've reached your goal.

What if you decide not to go for the Ultimate Test Bench? What if you get to Step Two and say, "Thanks, Ray. That's good enough for me." Then I say, "Great!" You'll still be going to the lab with more confidence than you have today, and that's the goal of this book.

1.4 The Seven Steps to the Ultimate Test Bench

There are seven steps from flat ground to the top of the verification staircase. Each step builds upon the ones before it and provides a small incremental improvement to your test bench designs. There are examples in all the chapters, and all the example code can be downloaded from www.fpgasimulation.com.

Step Zero is the base of the staircase. It assumes the following:

- You have access to a simulator
- You simulate your code just enough to see if the syntax is correct.

This is all you need to start climbing the steps. From Step Zero, we will walk through seven steps. You can add these steps to your design flow one at a time on a project-by-project basis. Each step gives you a stand-alone improvement over what you are doing today.

Here are the steps:

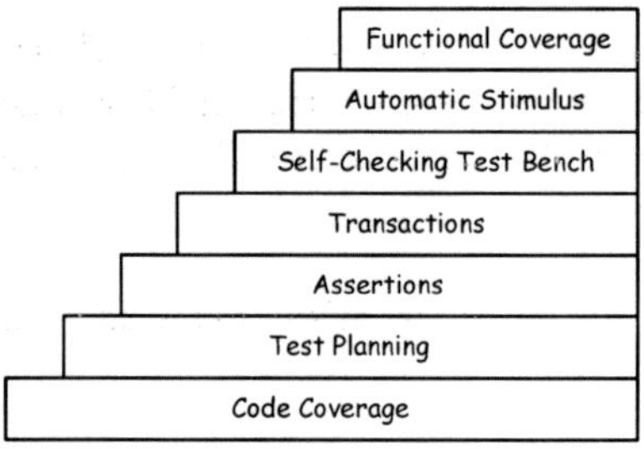

FIGURE 1-1. The Seven Steps to the Ultimate Test Bench

The goal is to add just one step to your next project, get comfortable with it, and add the next one. You can keep doing this until you decide that you are happy with your verification test bench or you reach the top.

Here are the seven steps:

1. **Code Coverage**—In this step you will take advantage of the code coverage features in your simulator to track how much of your RTL has been simulated.
2. **Test Planning**—When you turn on code coverage, you'll see how difficult it can be to achieve complete code coverage with an *ad hoc* testing approach. This step adds a test plan to your process to make verification go faster.
3. **Assertions**—Assertions catch bugs at their source, before they get to the output pins. Study after study has shown that they cut debug time in half. We'll learn how to add assertions easily to our project.
4. **Transaction-Level Simulation**—Transactions make it easy for you to create tests and check results. They encapsulate the complexity of communicating with a device.
5. **Self-Checking Test Bench**—Transactions make it easy to write tests, but we are limited in how many tests we can write if we need to use a manual process to check the results. The Self-Checking Test Bench solves this problem.

6. **Automatic Stimulus**—Once the test bench can check its output, it can start writing its own tests. Automatic Stimulus lets the test bench do the work.
7. **Functional Coverage**—Our test bench can now tell us whether it has tested every function in the design. Functional Coverage goes beyond code coverage to check that we've tested all our interesting data as well as RTL paths.

Your job now is to find where you are on this staircase and then start climbing.

1.5 Summary

In this chapter we looked at the lab debug problems that are raised by FPGAs that grow increasingly more complex. We saw that the solution is to simulate our FPGA designs thoroughly before we go to the lab.

There is no need to implement all the steps of FPGA simulation in a single project. We can take a step-by-step approach that improves our simulation test bench over time. We discussed the seven steps in the verification staircase.

Now its time to take the first step: code coverage.

2 Code Coverage

In This Chapter:

- *Statement Coverage*
- *Branch Coverage*
- *Condition Coverage*
- *Expression Coverage*
- *Finite State-Machine Coverage*

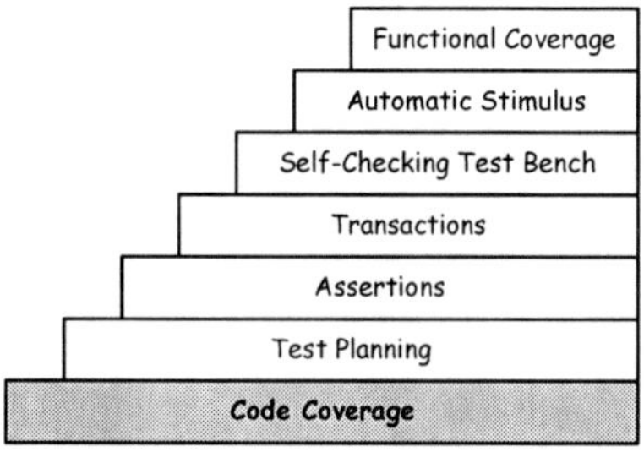

Of all the verification techniques, code coverage gives you the biggest bang for your buck. It is completely automatic. Turn it on, read the results, and you've got valuable information about what your tests are missing.

Code coverage is based on the idea that if you haven't simulated all the paths through your RTL, then you haven't fully tested it. It makes sure that you've touched every line and followed every branch in your code. Code coverage is built into all simulators from major EDA vendors. However, it is not built into the simulators that come with FPGA design tools. These simulators have limited functionality relative to their fully featured counterparts.

In all cases, you control code coverage when you compile and simulate. Generally there is a command line switch that you apply at compilation and another at simulation.

Code coverage does not require that you change your current approach to designing and simulating your FPGA. You simply run the tests you already have and look at the coverage results.

2.1 The Code Coverage Goal

Code coverage tells us how much of our design we've simulated. Simulators report this number as a percentage. The question is, what is the right percentage? How much coverage is enough?

I often hear engineers say that 95% code coverage is enough. That number doesn't work for me. I think of code coverage the way I think of a roof. How much of a roof is enough? If your roof is 95% waterproof would you be pleased with the results? The same is true of code coverage. Code coverage of 95% is not a sign that we've completed our simulation. It is just a sign that 5% of our code is pretty sure to have bugs.

At this point the alert reader points out that we can never really achieve 100% code coverage. There is always code that we can't or shouldn't simulate.

For example, let's say we are simulating a one-hot state machine, but we've put a default in our case statements so that an invalid state (e.g. two bits are set) transitions back to a known state. We cannot simulate that situation in our tests.

This is true, and since we can't simulate that code we should exclude it from the code coverage percentage. This allows us to achieve 100% code coverage with a set of exclusions. It's up to our team to review that list of exclusions and make sure that we are OK with not simulating them.

We discuss exclusions in detail after discussing the various forms of code coverage.

2.2 Our Example Design: The TinyCache State Machine

It is much easier to explain abstract concepts such as code coverage if we have an example design. In this chapter we use a state machine controller from a design called the TinyCache. This is a simple write-through cache that sits between a CPU and a memory.

The TinyCache contains a state machine, and that state machine has a process to figure out its next state. Here is the flow chart from that process:

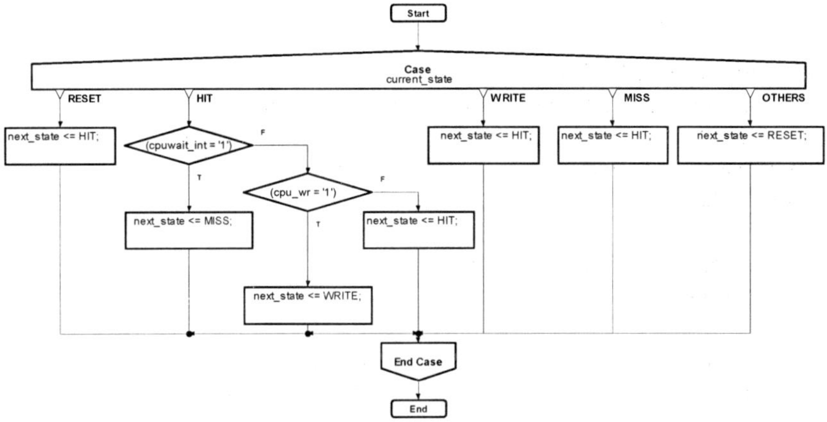

FIGURE 2-1. TinyCache Flow Chart[1]

As you can see, even this small design has many paths. We'll use the five forms of code coverage to make sure that we've simulated it fully.

2.3 Statement Coverage

Code coverage is a catch-all phrase that describes five ways of looking at a design and determining whether it has been tested:

1. Statement —Did I touch all executable statements?
2. Branch—Did I take all branches?
3. Condition—Did I take all branches for all reasons?
4. Expression—Did I test all parts of my concurrent assignments?
5. Finite State Machine—Did I touch all states and transitions?

1. This diagram was generated by HDL Designer from Mentor Graphics.

In this section we focus on statement coverage, the simplest form of code coverage. It checks to see whether we touched every executable statement.

A statement is a line of code that ends in a semicolon. Most statements are assignments or calls to tasks, procedures, or functions. Statement coverage gives us a gross view of how well we are covering our design.

As an example, we ran a test through the example design with statement coverage turned on. Here is the statement coverage report for the flow chart in Figure 2-1 on page 13:[2]

```
112                      ------------------------------------------
113                      nextstate_proc : PROCESS (
114                         cpu_wr,
115                         cpuwait_int,
116                         current_state
117                      )
118                      ------------------------------------------
119                      BEGIN
120                         CASE current_state IS
121                            WHEN RESET =>
122  1         4                  next_state <= HIT;
123                            WHEN HIT =>
124                               IF (cpuwait_int = '1') THEN
125  1        16                     next_state <= MISS;
126                               ELSIF (cpu_wr = '1') THEN
127  1   ***0***                     next_state <= WRITE;
128                               ELSE
129  1        10                     next_state <= HIT;
130                               END IF;
131                            WHEN WRITE =>
132  1   ***0***                  next_state <= HIT;
133                            WHEN MISS =>
134  1         8                  next_state <= HIT;
135                     -- coverage off
136                            WHEN OTHERS =>
137  1         E                  next_state <= RESET;
138                     -- coverage on
139                         END CASE;
140                      END PROCESS nextstate_proc;
```

FIGURE 2-2. Statement Coverage Report[3]

This report shows us that we never touched lines 127 or 132. From looking at the state names, we can infer that we've never tested the cache's write operation—because we never went to the `WRITE` state and we were never in the `WRITE` state.

2. All examples are available on www.fpgasimulation.com and are numbered to correspond to the sections of the book.
3. These reports are being generated by ModelSim from Mentor Graphics.

This is statement coverage, so we see that the simulator is looking only at statements that execute an operation. It ignores the `CASE` statement and the `IF` statements. We'll check these with other kinds of coverage.

You can see that we excluded the `OTHERS` case on line 136, by using the pragmas on lines 135 and 138. These lines are never simulated and should not be part of the code coverage metric.

2.4 Branch Coverage

Branch coverage tells us whether we've exercised all the possible branch directions in our design. As you can see in Figure 2-1 on page 13 even this simple design contains many paths.

Branch coverage looks at `IF` and `CASE` statements and makes sure that we've followed all the paths through them. It measures code coverage as the difference, in percent, between the possible number of branches that could have been taken and the number that were.

Here is the result of running branch coverage on our example design:

```
------------CASE Branch------------------------------------
    120                38     Count coming in to CASE
    121    1            4               WHEN RESET =>
    123    1           26               WHEN HIT =>
    131    1       ***0***              WHEN WRITE =>
    133    1            8               WHEN MISS =>
    136    1            E               WHEN OTHERS =>
Branch totals: 3 hits of 4 branches = 75.0%
```

FIGURE 2-3. Branch Coverage Report

The report shows that the `CASE` statement on line 120 has four covered branches but it touched only three of them. There is one uncounted branch because we excluded the `OTHERS` statement on line 136. In the 38 times through the case statement we touched `RESET` 4 times, `HIT` 26 times, and `MISS` 8 times. We never touched the `WRITE` path.

A little further down in the report we can see more-detailed branch coverage. Figure 2-1 on page 13 shows us that the `HIT` case has three paths through it. Here is the coverage report for the `IF` statement in the `HIT` case:

```
------------IF Branch-----------------------------------
   124              26     Count coming in to IF
   124    1         16                 IF (cpuwait_int = '1') THEN
   126    1     ***0***                ELSIF (cpu_wr = '1') THEN
                                            10      else
Branch totals: 2 hits of 3 branches = 66.7%
```

FIGURE 2-4. Branch Coverage in an IF Statement

The report shows us that there are three paths through this `IF` statement. The `IF` on line 124 can be true or false. The true case was handled 16 times and the false case was handled 10 times. The `ELSIF` case was never true and this is the missing path. We're starting to see a theme where writes have not been tested.

You would think that branch coverage and statement coverage would tell us the whole story. However, it is possible that we've tested whether we've taken a branch but we haven't tested all the reasons to take the branch. For that, we need another level of code coverage called condition coverage.

2.5 Condition Coverage

Consider an `if` statement that is true if A or B are true. In Verilog this would look like this:

```
if (A || B)
    inc_sum = sum+1;
```

Both statement coverage and branch coverage tell us whether this line was ever executed. Neither tells us that it was executed because `A` was true or because `B` was true. Condition `B` might never have been true and we wouldn't know. Condition coverage gives us that information.

We'll look at the Verilog version of our TinyCache to examine condition coverage. The TinyCache has a signal called `cpuwait`. That signal tells the CPU whether there is a cache miss. We generate the signal with an `always` block:

```
always @(key_ram[cache_address] or cpu_address or cpu_rd)
  if (((key_ram[cache_address] != cpu_address) ||
        (invalid [cache_address]) )&& cpu_rd)
     cpuwait = 1;
  else
     cpuwait = 0;
```

FIGURE 2-5. Using a Complex Condition in the TinyCache

This is a complicated `if` statement. It says that we have a cache miss if the address is not stored in the cache or if the cache has been reset. The `cpu_rd` signal qualifies both conditions. The condition summarizes to

```
(A or B) and C
```

with the following definitions for `A`, `B`, and `C`:

- `A`—Is this a cache miss?
- `B`—Is the cache entry invalid?
- `C`—Are we doing a read operation?

A cache miss happens when the `cpu_rd` signal (C) is high and either the address isn't in the cache (A) or the invalid bit for that cache entry is set (B.)

If we run branch coverage on this design, we'll see that we have both hits and misses. We've traveled down both paths of the `always` block. The question is, did we travel down them for all the possible reasons?

There are four condition combinations in this `if` statement:

1. We don't have a cache miss or an invalid entry, so `cpuwait` is 0.
2. We are not doing a read, so `cpuwait` is 0.
3. We are doing a read and this is a cache miss, so `cpuwait` is 1.
4. We are doing a read and this entry is invalid, so `cpuwait` is 1.

The other four combinations are covered by these four conditions. For example, if we have a cache miss and an invalid entry, then condition 3 or 4 handles it.

Here is the condition coverage report for our equation:

```
-----------------------------------Condition------------------------
Line        86 Stmt     1         (invalid [cache_address]) )&& cpu_rd)
Condition totals: 3 hits of 4 rows = 75.0%
Truth Table:              (key_ram[cache_address] != cpu_address)
                          |invalid[cache_address]
                          ||cpu_rd
                     hits |||(((key_ram[cache_address] != cpu_address)
                                 | invalid[cache_address]) & cpu_rd)
  Row    1:            31 00-0
  Row    2:            31 --00
  Row    3:        ***0*** 1-11
  Row    4:             8 -111
  unknown:              0
```

FIGURE 2-6. Condition Coverage for the `cpuwait` Branch

Condition coverage shows us that we've covered only three of the four possible combinations. We missed the combination where we are doing a read and we don't have the right address. In fact, the test registered eight cache misses and they were all due to the invalid bit being set on the cache entry. Even though statement coverage tells you that you've tested the cache miss statements, and branch coverage says you've had hits and misses, you know that you still have some more testing to do.

2.6 Expression Coverage

RTL designs contain two classes of statements: procedural statements and concurrent assignments. Procedural statements are stored in processes and run step by step.

Concurrent assignments sit outside the processes and automatically run whenever one of the variables on their right-hand side changes. In Verilog we use the `assign` keyword to create concurrent assignments. Expression coverage makes sure that we've tested concurrent assignments completely.

In our last example we used a Verilog `always` block to create a `cpuwait` signal in our cache. This signal can also be created by using a concurrent assignment. Here is the Verilog that does the job:

```
assign cpuwait =  ((key_ram[cache_address] != cpu_address) ||
                            (invalid [cache_address]) )&& cpu_rd;
```

FIGURE 2-7. Complex Continuous Assignment in TinyCache

When we simulate this design with expression coverage, we get the following results:

```
-----------------------------------Expression------------------------
Line        84 Stmt     1                         (invalid [cache_addr
Expression totals: 3 hits of 4 rows = 75.0%
Truth Table:               (key_ram[cache_address] != cpu_address)
                           |invalid[cache_address]
                           ||cpu_rd
                     hits |||(((key_ram[cache_address] != cpu_address)
                                    | invalid[cache_address]) & cpu_rd)
  Row    1:            31 00-0
  Row    2:            33 --00
  Row    3:      ***0*** 1-11
  Row    4:             8 -111
  unknown:              1
```

FIGURE 2-8. Expression Coverage in TinyCache

Expression coverage works just like condition coverage in that it calculates the minimum set of possible results and then lets you know whether you've covered all of them. Once again, we missed the case where there is a cache miss as a result of not having the right entry in the cache.

2.7 Finite State-Machine Coverage

Some simulators add another level of abstraction to code coverage by checking the coverage in the state machines. Finite state-machine (FSM) coverage makes sure that we've touched every state in the state machine and followed every transition.

This information is available in other forms. The branch and condition coverage capture missing transitions, and the statement coverage captures missing states. Still, it is easier to debug our tests if we think of these misses in terms of our state machines.

Consider the following TinyCache state machine diagram:[4]

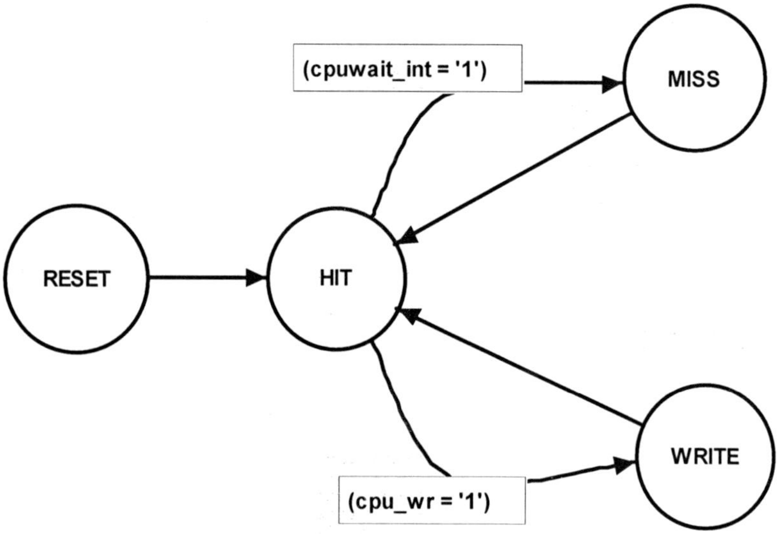

FIGURE 2-9. TinyCache State Machine

The state machine has four states. One of them is the `RESET` state, and the state machine jumps to it upon a reset. Otherwise, the state machine moves between the `HIT` state and the `MISS` and `WRITE` states. Implied in this diagram is the fact that we sit in the `HIT` state if neither of the exit conditions is true. When the cache is in this `HIT` state, it just responds like a RAM.

The cache generates the `cpuwait_int` signal by comparing the address on the address bus to an internal table of stored addresses. If the address on the address

4. This diagram was generated with HDL Designer from Mentor Graphics.

bus is in the table, then we have a hit. Otherwise, the `cpuwait_int` signal goes high and the cache implements a cache miss.

The CPU raises the `cpu_wr` signal when it requests a write operation.

When we run our test we get the following FSM coverage report from Model-Sim:

```
25      Covered States :
26      ----------------
27                    State            Hit_count
28                    -----            ---------
29                    reset                    1
30                      hit                   42
31                     miss                    8
32      Covered Transitions :
33      ---------------------
34                 Trans_ID           Transition                Hit_count
35                 --------           ----------                ---------
36                        0         reset -> hit                        1
37                        2          hit -> miss                        8
38                        4           hit -> hit                       32
39                        5         hit -> reset                        1
40                        8          miss -> hit                        8
41      Uncovered States :
42      ------------------
43                    State
44                    -----
45                    write
46      Uncovered Transitions :
47      -----------------------
48                 Trans_ID           Transition
49                 --------           ----------
50                        1       reset -> reset
51                        3         hit -> write
52                        6         write -> hit
53                        7       write -> reset
54                        9        miss -> reset
```

FIGURE 2-10. FSM Coverage Report from ModelSim

The report is broken into two sections: what we've covered and what we haven't. On line 29 we see that we've covered the `RESET`, `HIT`, and `MISS` states. We've missed the `WRITE` state.

This lends new insight to the hints we've seen throughout this chapter. You can see in the statement coverage that we never touched the statement that set the memory write signal to 1. In the branch coverage we saw that `cpu_wr` signal was never high. The reason behind both of these clues becomes clear here in FSM coverage. We never had a write test.

We see more evidence of this in the transition coverage. We covered all the transitions associated with hits and misses, but none of the transitions associated with the `WRITE` state.

The transitions to and from the `RESET` state pose an interesting question. The `reset` signal can become active in any state. We've seen it become active only in the `HIT` state, because this is the basic waiting state in the state machine. Do we want to test all the possible resets? If we do, then we'll need to modify our test bench to throw resets randomly into the mix.

FSM coverage gives us the last piece of the puzzle for a complete view of our testing. If we have not tested all our states and transitions, then we know we haven't tested our state machines fully.

2.8 Toggle Coverage

I don't recommend using toggle coverage for RTL design simulation, because toggle coverages wasn't created for RTL design analysis. It was created to support chip testers in a chip fab.

Chip testers use signal vectors to make sure that all the dies on a wafer are good. They do this by applying vectors to the chips and looking at the output. The chip testers are looking for, among other things, flaws that cause a node in the chip to be stuck at logic one or logic zero. Toggle coverage allowed engineers to simulate all the vectors in a design and see whether they had achieved 100% coverage.

However, this application doesn't apply to RTL. While you can run a toggle coverage report on your RTL, you'll probably waste a lot of time chasing false holes in toggle coverage—rather than simulating your chip.

2.9 Exclusions

Exclusions allow us to designate portions of the code that should never be tested. Exclusions should be used carefully and should be reviewed. It's easy to get to 100% coverage if we exclude all the hard cases.

Exclusions come in several forms:

- **Block Exclusion**—In this case, you pick a block of code and tell the simulator not to count it when it creates the report. This block of code should never be executed. You might use this with debug logic that you've put into your design.
- **Default Exclusion**—Some simulators, such as ModelSim, have options that tell the simulator to exclude all `default` and `OTHERS` options in `case` statements.
- **Conditional and Expression Row Exclusion**—Branch coverage and expression coverage may find combinations of signals that should never exist. For example, there may be a mode bit in a design that never changes. You'll need to exclude the rows where that bit is zero.
- **FSM State and Transition Exclusion**—As we saw in the case of our `RESET` transition in Section 2.7, we may want to ignore certain unused state transitions or certain error states.

Exclusions are an important part of your code coverage strategy. 100% coverage is the easiest to measure and use as a milestone. You can attain it only if you exclude the proper areas of the design. If you don't use exclusions, then you'll have to pick some number, say 95%, and claim that this represents complete verification. Perhaps it does, or perhaps it doesn't. So rather than create some squishy goal, set the coverage goal at 100% and exclude untestable lines explicitly.

2.10 Summary

We started this discussion with a new schedule milestone: Verification Complete. We said that this milestone would need an easy-to-use measure of completion. Code coverage supplies the simple, measurable, milestone you need to be sure that you've completed your chip's verification. Achieving 100% code coverage reduces your time in the lab.

Avoid the temptation to accept less than 100% coverage as your Verification Complete milestone. If you are getting less than 100% code coverage, there can only be three reasons:

- You haven't written enough tests to test your design completely. Write them.
- You haven't excluded code and conditions that should never happen. Exclude them.
- You have dead code in your design. Delete it.

Any one of these three problems can create difficulties in the lab. Be sure to fix them before you waste time.

The fastest way to write the tests you need is to create a test plan, and that's the subject of Step Two: Code Coverage.

3 Test Planning

In This Chapter:

- *The Virtue of Planning*
- *Capturing DUT Functionality*
- *Describing Stimulus and Response*
- *Creating the Test List*

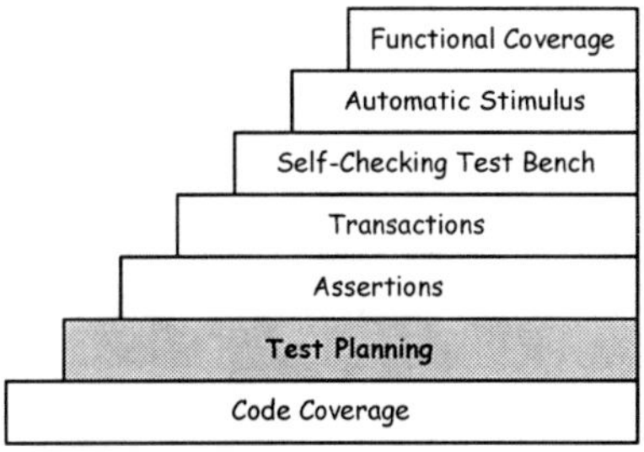

In Step One, we turned on code coverage. Now we know exactly how much coverage our current testing has been providing. Unfortunately, this information usually causes people to go through the five stages of grief:

1. Denial—That's got to be wrong!
2. Anger—This is ridiculous! Nobody could write this many tests!
3. Bargaining—If I simulate the hard stuff, the easy stuff will have no bugs.
4. Depression—Oh, #$(&@!
5. Acceptance—I guess I had better start writing tests.

We start writing our test plan after the point of acceptance. The test plan outlines how we achieve the code coverage we need for a successful lab debug. We write the test plan to fulfill a simple rule:

All RTL statements must be simulated before the chip goes to the lab.

This isn't to say that we need to test every path and every last corner case of our algorithm exhaustively. This isn't an ASIC design, where a bug could cost a million dollars. However, it does accept the fact that unsimulated RTL statements are very likely to have bugs, each of which costs anywhere from three hours to three weeks to fix. We need a plan to test all the code *before* we hit the lab.

The approach we'll outline in this chapter organizes our thoughts for creating tests with any test bench approach. It also delivers the groundwork we'll need when we start looking at more advanced SystemVerilog-based test benches.

3.1 The Virtue of Planning

At this point, the alert reader will raise the following question: "Why is planning Step Two in this book? Shouldn't we do planning first?"

The answer is first, "Good Question," and second, "Yes." We should do planning first. Remember, though, that the seven steps of verification are a plan to help your team adopt new simulation techniques. They are not a verification plan. Step One, "Turn on Code Coverage," is the easiest step for a team that isn't doing any simulation.

The goal was to create a first step that didn't modify the team's work flow. The team just compiles with new switches and looks at the coverage number. Writing a test plan requires the team to take new steps, and it was unlikely that the team would be motivated until it saw its coverage numbers.

There are those who say that creating a test plan is just paperwork. They contend that writing down what you're going to do before you do it adds no value to the process. Let's consider this idea in terms of tree houses.

When I was young (and phones had dials) my friends and I decided to build a tree house. Building a tree house required three things:

1. A tree
2. Some wood
3. Some nails and a hammer

Our tree house construction project worked much like an FPGA project with no test planning. We stood around the tree and looked at the wood. Then we argued about which piece of wood should go where and finally we started building.

If we were lucky there was only one hammer, so that only one piece of wood went up at a time. If there were more hammers, then wood could go up following a variety of visions. In the end, we would make it work by cobbling together the scraps and we'd have a rickety platform nailed to a tree—a tree house.

This process of development was fine for a tree house. It wouldn't work for a real house. Throwing a bunch of contractors together with a pile of wood and asking them to start building is a recipe for disaster. You would get a house but you wouldn't want to live in it. In fact, you'd probably need to tear it down and start over.

The same is true of FPGAs that are verified without a test plan. You can put together a rickety test bench with simulation scripts and verbal promises that "Oh yeah, I simulated that," but you wouldn't want to live with the results. You certainly wouldn't want that FPGA in your airplane.

Planning is better than not planning in almost every aspect of human achievement. Yet it can be difficult to get a design team to sit down and create a plan. This is often because they believe that planning is extra work. Even though it isn't.

Team's inevitably do all the work in a test plan in order to ship the FPGA. Most teams don't notice that they are doing the work, because it gets spread out over the project. Therefore, the test planning work is hard to measure and folks don't realize they are doing it. On the other hand, because they are not measuring their plan completion, these teams don't know whether they've created a complete plan or when they are done.

By compressing test planning into one clearly defined project milestone, the team gets test planning finished and moves on.

We do three things to create a test plan:

1. Capture device under test (DUT) functionality.
2. Describe stimulus and response.
3. Create the test list.

If we've done these three things, we have written a complete test plan.

3.2 Capture DUT Functionality

Capturing the functionality in a complicated chip is a difficult task. One could sit down with the specifications and simply copy the behavior into the test plan. Nevertheless, this won't help us when we need to describe how we'll stimulate the design or check the results. Instead, we need a more systematic approach. We need to work from the pins in.

All DUTs have data inputs and data outputs. These input and output paths are called *channels*. We are going to write our test plan by looking at the channels and how they behave. Then we are going to use the channels to describe the DUT's behavior.

3.2.1 Listing the DUT Channels

A channel is a way of getting data into a block or pulling it out. We use the channels to define the DUT's behavior clearly, so that we can write stimulus and check results. Channels have the following features:

- **Channels are unidirectional**—They either put data in or take data out. We describe a bidirectional bus as two channels: one in and one out.
- **Channels are independent of pins**—We look at pins and ports later in the process. For now, we just worry about the data flow.
- **Channels focus only on the DUT**—There are designs where our DUT can communicate with many other blocks through a shared bus. We do not create channels for every possible target. Instead, we create a single channel that connects the DUT to the bus.

Using channels to define all the data flow into and out of a DUT gives us a foundation to build a test plan and eventually a test bench. Let's look at a small example.

The TinyCache is a write-through cache that sits between a CPU and a large memory. The cache has four channels.

TABLE 3-1. TinyCache Channels

Channel	Description
CPU2Cache	Read and write requests from the CPU
Cache2CPU	Responses to CPU read requests
Cache2Mem	Read and write requests from the cache to the memory due to cache misses or write requests
Mem2Cache	Responses to cache-read requests

This design has four channels because it sits on two bidirectional ports. The channel names are just for documentation. We follow the convention of naming them with the source and destination by using the number 2, for example `Cache2Mem`.

3.2.2 Channel Transactions

We understand complicated designs by breaking them down into clearly defined pieces and describing how those pieces work together. We started this process by creating a list of the channels into and out of the chip. Now we need to understand these channels more deeply.

Transactions are the actions available on a channel. Some simple channels, such as memory channels, have only read and write transactions. Other more-complicated channels may have block transfers or read-modify-write transfers. There may be different transactions, based upon the number of cycles or the width of the transaction.

Transactions are not operations. Operations are pieces of functionality that a chip can execute. For example, an ALU may have four operations (ADD, AND, XOR, MULT,) but all these are requested by means of the same kind of transaction.

A test plan's primary job is communication. Therefore, we want to describe the transactions as clearly as possible. We do this by writing the transactions as if they were software function calls. We give them an operation name and a set of arguments. For example, we could create a transaction on a memory channel that looks like this:

```
read(address)
```

This is a unidirectional transaction. It tells the test bench to do a read on the DUT's input channel. The result is another transaction on the DUT's output channel.

The function call format serves two purposes:

1. It is very clear. We can see exactly what this transaction is supposed to do and the data it needs to do it.
2. We can use the format later on when we build procedures that implement the transaction.

We capture the transactions in our test plan by making a table that shows all the channels in our DUT and all the transactions that each channel can implement. This needs to be an exhaustive list of everything the channel can do.

As an example, here are the transactions for the TinyCache's four channels:

TABLE 3-2. Transactions for TinyCache Channels

CPU2Cache	Cache2CPU	Cache2Mem	Mem2Cache
read(address)	result(data,hitormiss)	read(address)	result(data)
write(address,data)	result(data,hitormiss)	write(address,data)	

Notice that we've captured the fact that the TinyCache can return data as either a cache hit (which takes one cycle) or a cache miss (which takes multiple cycles while the cache reads the main memory). Though we are not concerned about cycles yet, we know we will be concerned in the future and we are planning ahead.

3.2.3 Functionality List

All DUT functionality can be broken down into a series of transactions. We put data into one channel and we get data out of another channel. When we write tests, we use this fact to write stimulus and check responses.

It is possible to have transactions that enter a chip and change its state. A simple example is a UART that converts parallel data to serial data. We can write to a register to change the UART's baud rate. This transaction does not produce an output, but all later output transactions behave differently.

When we have complexities such as these, it is even more important to create a test plan. We need to know what correct behavior looks like, and if we don't clearly define it now we'll have to define it eventually. Defining the correct behavior as early as possible beats standing in the lab, looking at a logic analyzer, and wondering, "Is that a bug?"

Writing a test plan for a complicated design also helps on single-engineer projects where we write the design and all the tests. By clearly defining the interfaces, we get a sharper picture of what the design is supposed to do and we create code with fewer bugs.

We use the channels in a table to describe each of our DUT's behaviors. The table columns represent the channels and the rows represent time. We don't think in terms of clock cycles, but rather in terms of which transactions go into the DUT and which ones come out.

Here is the functionality in the TinyCache captured in a table:

TABLE 3-3. TinyCache Functions in Terms of Transactions

CPU2Cache	Cache2CPU	Cache2Mem	Mem2Cache
Read with Cache Hit			
read(address)			
	cachehit(data)		
Read with Cache Miss			
read(address)			
		read(address)	
			result(data)
	cachemiss(data)		
Write			
write(address,data)			
		write(address,data)	

This diagram shows us how the TinyCache works in terms of channels and transactions. We read the chart downward line by line. All three operations start with a request on the CPU2Cache channel. In the case of the cache hit functionality, the read transaction on the CPU2Cache is immediately followed by a result transaction on the Cache2CPU channel that shows us that we've had a cache hit.

The cache miss and write operations generated additional transactions on the Cache2Mem and Mem2Cache channels. When we test our cache we need to make sure that this downstream behavior occurred.

3.3 Describing Stimulus and Response

In this section we describe the input and output transactions in terms of pin behavior on all the channels. We use the input transactions to create stimulus procedures that drive data into the DUT. We use the output transactions to recognize the data coming out of DUT, so we can make sure it is correct.

3.3.1 Defining Stimulus Procedures

When we defined our transactions we used a format that looked like a function call. We did this for clarity and also so that we could create a library of stimulus procedures.[1] These procedures drive our DUT's pins and put data into our chip.

Our test plan should look like documentation for a programming interface. Each procedure has a name, a list of arguments, and a waveform to show what the procedure is doing on the interface.

For example, here is the documentation for the `read()` transaction in the CPU2Cache channel:

```
read(address)

* address: A memory address

This transaction executes a read function on the TinyCache as
if the CPU were making the request. The address is passed to
the cache on the address bus.

If the cpuwait signal goes high during the read, then there
has been a cache miss and the task waits.
```

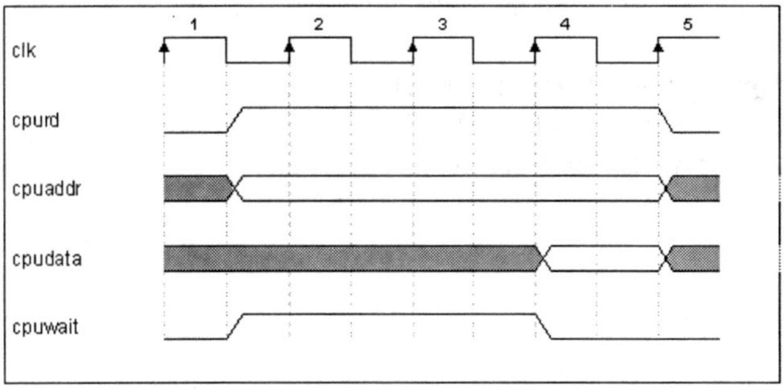

FIGURE 3-1. Example Transaction Description

1. VHDL and Verilog both allow subroutines that can drive signals and wait for clocks. VHDL calls these "procedures," whereas Verilog calls these "tasks." I use the VHDL term in this book, but when I say "procedures" I mean "procedures or tasks."

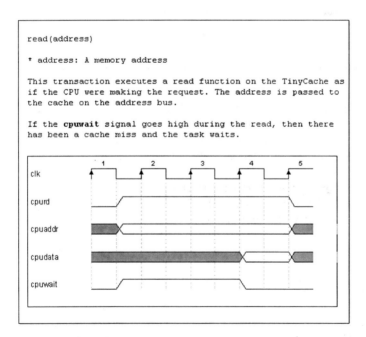

This is all the information our test writers need to write their tests. They just call this procedure and the test bench executes a write function.

Notice that this description doesn't discuss how the data comes back from the cache. That is the output channel's job. This description shows only how the chip is stimulated. That is not to say that we don't look at outputs here. The `cpuwait` signal is an output from the cache, but we only use it as part of our input transaction.

At this point in our test plan, we create a section that consists of nothing but the transaction procedure documentation. This leaves us the job of describing the output transactions.

3.3.2 Describing Output Transactions

We create a test bench by using the input channels to stimulate the DUT and the output channels to test the DUT. The transactions our DUT creates on the output channels need to match the ones we expect from our functional specification.

There are several ways to check the transactions on an output channel:

- **Look at the waveforms**—This is a time-honored but error-prone way of checking output.
- **Add transaction information to the waveform**—This feature is available from several simulators.
- **Capture transactions and convert them to a string**—In this case we create a state machine that recognizes transactions and writes a string to the screen or to a file.

The technique we use to view the output transactions depends upon our implementation methodology. However, regardless of the technique we use to view the output transactions, they all must be described.

We do this by going through each output channel and listing the transactions available on it. Then we provide a waveform and description for each transaction. Our test bench developers can use these descriptions to implement a strategy for viewing outputs.

The TinyCache has two output channels: Cache2CPU and Cache2Mem. The Cache2Mem channel has two transactions: `read(address)` and `write(address,data)`. Both of these place an address on the interface and raise the read or write signal to the memory.

We describe these transactions the same way we described the input transactions. However, we don't write tests with these transactions. Instead, we use them to check the results on the output channel.[2] The output transactions we get need to match the ones in our functional specification.

We create a transaction description for every transaction in the output channel by using our functional call format. Here is the description for the `read` transaction from the `CACHE2MEM` channel in Table 3-2 on page 30:

```
read(address)

* address: The address we read from
```

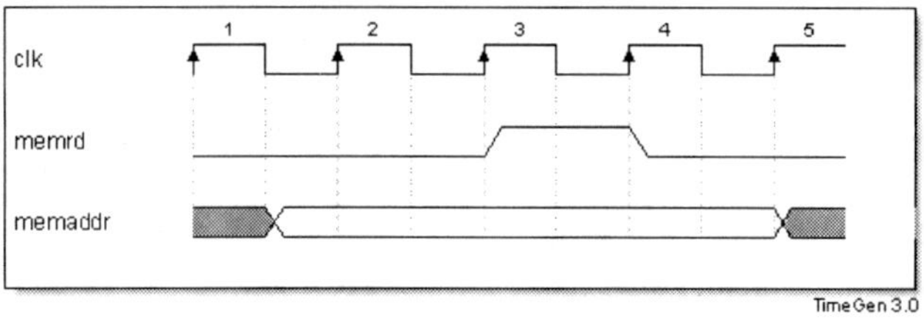

FIGURE 3-2. Output Transaction Description

Notice that we don't bother describing the data bus in this description. This is because the data is not part of this output transaction from the cache to the memory. The `memaddr` bus is valid for several clocks before the `memrd` signal goes high, because that address came straight through from the CPU read. As long as we see the cache raise the `memrd` signal when we expect it, we know that we are looking at a `read` transaction.

2. In Step Four, Transactions, we implement a technique that does this with SystemVerilog.

3.4 Create the Test List

Now we combine the channels and transactions with the complete functional list to create a test list. The test list describes the tests in terms of the input transactions and the expected output transactions.

This list does not worry about the details of how these tests are implemented or checked. That can change from test bench to test bench or from team to team. Instead, this list is focused on understanding how much work needs to be done to test the design.

Our goal is 100% code coverage. The test list helps us understand which functionality we need to exercise to achieve our goal. As we'll see, we may not need explicit tests for all functions to get to 100% code coverage.

The test list is a table that describes each test in terms of the stimulus procedures that we'll call to drive the DUT, as well as the transactions we expect to see on the output channels.

TABLE 3-4. Test List for the TinyCache

Test Name	CPU2Cache	Cache2CPU	Cache2Mem
rEmptyCacheMiss	read(A)[a]	result(data,MISS)	read(A)
rAddressNotInKey	reset read(A) read(A+page)	result(data,MISS) result(data,MISS)	read(A) read(A+page)
rCacheHit	reset read(A) read(A)	result(data,MISS) result(data,HIT)	read(A)
wCacheHit	write(A,data) read(A)	result(data,HIT)	write(A,data)
wReadMem	write(A,data) reset read(A)	result(data,MISS)	write(A,data) read(A)
wSameCacheLoc	write(A,data) write(A+page,data) read(A)	result(data,MISS)	write(A,data) write(A+page,data) read(A).

a. All tests assume a reset before they run.

The first column describes the test names. These tests are implemented differently in different test benches. For example, in a VHDL test bench we may have an entity called `stimulus` and then a series of architectures named after each test in this list. In Verilog we might simply have modules with the names.

However we implement the tests, it is essential to list all the stimulus procedures the test writer can call and the expected transaction responses. This allows us to write tests quickly.

The test list can also tell us whether we are oververifying. The TinyCache test list above was written from the TinyCache functional specification. If we have a requirement to test all functionality fully, we would need an exhaustive list—one for each function. However, code coverage may show that we don't need a test for every function.

If our goal is 100% code coverage, this list might be overkill. Consider the test `rCacheHit`. That test uses a cache miss to load the cache location and then generates a cache hit. That means that this single test gives us code coverage on our cache-miss code as well as on our cache-hit code. Therefore, we really don't need the test `rEmptyCacheMiss`.

This is the kind of optimization we can do if we create a test plan before we sit down and start writing tests. Once we have this plan, we have a complete picture of what it takes to test our DUT, and we can use that information to create an efficient test bench and a minimal set of tests.

3.5 Summary

In this chapter we saw that it can be difficult to reach 100% code coverage with an *ad hoc* testing approach. A test plan solves this problem.

We learned how to create a test plan in terms of input and output channels and transactions. We walked through the process of creating a test plan:

- Capture all functionality in terms of channels and transactions.
- Define the transactions in terms of waveforms and create stimulus procedures for test writing.
- Create a list of tests in terms of input procedures and expected output transactions.

We saw that creating a test plan in this way gives us a clear picture of the type of test bench we need to create. It also gives us a clear picture of the amount of work involved in reaching 100% code coverage. Test planning gives us the ability to dig quickly into the corners of our design.

However, many times digging into the corners simply finds more bugs, and these make the project schedules slip out. There are many ways to handle this problem. One is to stop simulating so that you won't find any more bugs (this has actually been done on some projects). This is an effective, but alarming, approach.

The solution, of course, is not to cover our eyes and chant, "There are no bugs. There are no bugs." What we really need to do is fix bugs more quickly. In the next chapter we'll work with a design technique that consistently cuts debug times in half. We're going to Step Three: Assertions.

4 Introduction to Assertions

- *Implementing Immediate Assertions*
- *Making Immediate Assertions with the* `assert` *Statement*
- *Using Multicycle Assertions*
- *How to Place Assertions*

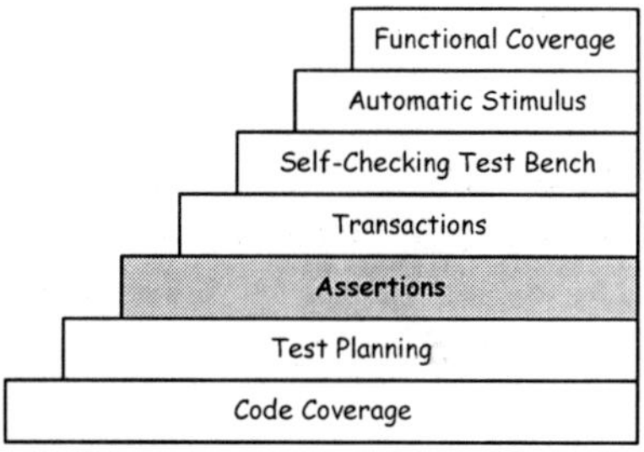

Most debugging sessions work the same way. We run our design, either in simulation or in the lab, and find the wrong data coming out of it. Then we scratch our heads and try to work our way backwards to the point in time when the design went bad.

The problem with this method is that we need to wait for the bad data to appear on the output pins. By that point, who knows where that data has been? Wouldn't it be nice if the design would tell us about the bad data as soon as it was generated? Then we wouldn't need to wait for it to work its way through the design.

Assertions allow us to do just that. They are checkers that run in our design as it operates. If our design does something dumb, assertions catch it at that moment and notify us. Assertions dramatically shorten our debug time, because we don't need to work backwards from the output pins. In fact, study after study[1,2] has shown that putting assertions into a design reduces the debug time by 50% or more.

1. [Abarbanel et al. 2000] Y. Abarbanel, I. Beer, L. Gluhovsky, S. Keidar, Y. Wolfsthal, "FoCs--Automatic Generation of Simulation Checkers from Formal Specifications," Proc. Computer-Aided Verification, 12th International Conference, CAV 2000, pp. 414–427, July 15–19, 2000.
2. [Turumella and Sharma 2008] B. Turumella, M. Sharma, "Assertion Based Verification of a 32-Thread SPARC CMT Microprocessor," Proc. Design Automation Conference, DAC 2008, June 8–13, 2008.

Assertions are not limited to simulation. We can put assertions into our synthesized design so that our chip raises a signal on a logic analyzer or lights an LED if an assertion fires in the lab.

In this chapter, we learn about assertion terminology and where to put assertions in our designs. In Chapter 5, we learn about the Open Verification Library, which provides ready-to-use assertions. In Chapter 9, we'll see how to synthesize assertions so they can help us debug in the lab.

4.1 Introduction to Assertions

We've all written assertions without knowing it. Our first assertion is usually a `printf` statement in a C program. When we added a `printf` that said, "I am here," we were creating an assertion. If we got fancy we could put a condition before the `printf` and catch an error. When we print to the screen we say the assertion *fired.*

There are two kinds of assertions, classified by how they relate to multiple clock cycles:

- **Immediate Assertions**—These assertions just look at data on a single clock cycle and fire if there is bad data on any cycle. An assertion that checks to make sure that a `read` and `write` signal on a RAM are not high at the same time is an immediate assertion.
- **Multicycle Assertions**—These assertions look at signals over several clock cycles and make sure that the signals follow the correct protocol. For example, a RAM might require that the address remain unchanged as long as the `write` signal is high. We would check this behavior with a multicycle assertion.

Immediate assertions can easily be implemented with VHDL and Verilog, while multicycle assertions cannot.

4.2 Implementing Immediate Assertions

Immediate assertions look at the data in our design and decide "immediately" whether to fire. They are useful for checking that signals are in a self-consistent state.

For example, suppose I have a three-bit counter that should never be incremented if it contains a `3'h7`[3] because we don't want it to roll over. Here is an immediate assertion that checks this behavior. The counter has reset, load, and increment functionality:

```
1  module threebitcounter (input clk, input rst,
2      input ld, input inc,
3      input [2:0] data_in,
4      output reg [2:0] data_out);
5
6    always @(posedge clk)
7      if (rst)
8        data_out = 0;
9      else
10       if (ld)
11         data_out = data_in;
12       else
13         if (inc) begin
14           // synthesis translate_off
15           if (data_out == 3'h7)
16             $display ("ERROR: Overflow");
17           // synthesis translate_on
18           data_out = data_out + 1;
19           end
20 endmodule // threebitcounter
```

FIGURE 4-1. Three-Bit Counter with Simple Assertion

- **Lines 7–8**—We reset if the `rst` signal is high.
- **Lines 10–11**—We load data into the counter if the `ld` signal is high. Notice that the load signal has priority over the increment signal. Because this behavior is defined, we don't care if both signals are high at the same time, so we don't have an assertion for that.
- **Lines 14–19**—This is the increment code, with the assertion implemented as the `if` statement on lines 15 and 16. There is nothing mysterious about this assertion. It is just a `$display` controlled by an `if` statement. We'll see how to make this code more compact later in this chapter.

3. This numerical format comes from Verilog. It means that this is a hex value that is three bits wide. You could also write this as 3'b111, which is a binary value that is three bits wide.

When we simulate our example we see the error:

```
# Loading work.threebitcounter
#                        0 data_out: x
#                       10 data_out: 0
#                       30 data_out: 1
#                       50 data_out: 2
#                       70 data_out: 3
#                       90 data_out: 4
#                      110 data_out: 5
#                      130 data_out: 6
#                      150 data_out: 7
# ERROR: Overflow
#                      170 data_out: 0
#                      190 data_out: 1
#                      210 data_out: 2
# ** Note: $finish     : top.v(25)
#    Time: 220 ns  Iteration: 0  Instance: /tb_top
```

FIGURE 4-2. Results of Simple Assertion

Immediate assertions are easy to add to your code and they are very useful. There are design teams that use nothing but immediate assertions and see tremendous value.

Immediate assertions are so valuable that SystemVerilog and VHDL have an `assert` statement that we'll use to replace the `if` and `$display` in this example.

4.3 Immediate Assertions with the `assert` Statement

The `assert` statement works the same way in SystemVerilog and VHDL. You pass it an expression that should evaluate to be true. If it is not true, then the `assert` statement creates an error message.

4.3.1 The SystemVerilog `assert`

The SystemVerilog `assert` statement is a procedural statement that can go into an `always` block or an `initial` block. The statement tests its argument whenever it is executed, and if the argument returns 0, X or Z, the `assert` statement generates an error message.

You can optionally provide blocks of code to run when `assert` passes or fails. If you don't provide blocks of code, then the assert writes to the screen only when it fails.

When we replace the `if` and `$display` in Figure 4-1 on page 41 with an assert statement, we get the following code:

```
module threebitcounter (input clk, input rst,
    input ld, input inc,
    input [2:0] data_in,
    output reg [2:0] data_out);

    always @(posedge clk)
        if (rst)
            data_out = 0;
        else
            if (ld)
                data_out = data_in;
            else if (inc)begin
             // synthesis translate_off
                assert (data_out < 3'h7) ;
             // synthesis translate_on
                data_out = data_out + 1;
            end
endmodule // threebitcounter
```

FIGURE 4-3. Three-Bit Counter with `assert` Statement

The `assert` statement gives us cleaner code than the `if` statement in Figure 4-1. This is an important point. Having `if` statements all through your code to check for values can make the code difficult to read. It's easier to have just a single `assert` line take care of your checks. We can also use the `assert` statement to check the status output of functions as we call them.

If the `assert` statement fires, then a SystemVerilog simulator calls the system task `$error`. The `$error` system task flags an error in the simulation log and writes information to the screen to identify the error.

The *SystemVerilog Language Reference Manual* (LRM) requires that all SystemVerilog simulators write the following information to the screen when `$error` is called:

- The file name and line number of the assert statement.
- The scope of the assertion.

Simulators do other things with this information as well. For example, ModelSim makes a mark in the waveform viewer to show that an error message was printed to the log at a certain time.

In our example, the error looks like this:

```
#                        0 data_out: x
#                       10 data_out: 0
#                       30 data_out: 1
#                       50 data_out: 2
#                       70 data_out: 3
#                       90 data_out: 4
#                      110 data_out: 5
#                      130 data_out: 6
#                      150 data_out: 7
# ** Error: Assertion error.
#    Time: 170 ns  Scope: tb_top.DUT File: threebitcounter.v Line: 14
#                      170 data_out: 0
#                      190 data_out: 1
#                      210 data_out: 2
```

FIGURE 4-4. Error Caught with `assert`

You can replace `assert`'s automatic behavior with your own code by providing the actions that the simulator does when the assertion passes or fails. Generally speaking, we are interested only in failure. In those cases we use the `else` keyword to designate the action.

```
10            if (ld)
11               data_out = data_in;
12            else if (inc)begin
13             // synthesis translate_off
14               assert (data_out < 3'h7) else
15                  $warning($realtime," Scope:%m  Counter Overflow");
16             // synthesis translate_on
17               data_out = data_out + 1;
18            end
```

FIGURE 4-5. Supplying an `assert` Error Message

In this example, we created our own message with the `else` keyword and `$warning`. This lets us control the severity of the assertion error.

4.3.2 Creating an assert in Verilog

The assert statement is an innovation in the Verilog world that was supplied by SystemVerilog. There is no assert statement in Verilog. Not to worry, we can make our own using a Verilog task.

```
    task assert;
      input condition;
      begin
         if (!condition)
            $display("Assertion Error");
      end
    endtask
```

FIGURE 4-6. Verilog '95 assert

This task implements the assert functionality by checking for a condition and then displaying an error if the condition is false. This approach raises the question of where one puts the task definition. We'll address this issue in Section 4.5 on page 47.

4.3.3 VHDL assert and report Statements

We can implement the same assertion functionality in VHDL. The assert statement in VHDL works the same as it does in SystemVerilog.[4] The statement takes a boolean expression as its argument and then prints an error message to the simulation screen and log if the boolean expression is false. The VHDL assert statement has two optional parts:

```
assert <condition>
   [report "message"]
   [severity <note |warning | error | failure>]
```

All simulators display some sort of information from this statement, even if they format it differently.

4. Here, the VHDL fan points out that VHDL had it first.

Here is our three-bit counter in VHDL with an assert to check for overflow:

```
36      process0_proc : PROCESS (clk)
37         BEGIN
38            IF (clk'EVENT AND clk = '1') THEN
39          -- Synchronous Reset
40          IF (rst = '1') THEN
41              data_out <= "000" ;
42          ELSE
43             IF ld = '1' THEN
44                data_out <= data_in;
45             ELSIF inc = '1' THEN
46                -- synthesis translate_off
47                assert (data_out /= "111")
48                  report "Oi! Counter Overflow"
49                  severity warning;
50                -- synthesis translate_on
51                data_out <= std_logic_vector(data_out + "001");
52             END IF;
53          END IF;
54            END IF;
55      END PROCESS process0_proc;
```

FIGURE 4-7. VHDL `assert` Statement

The `assert` statement on line 47 works the same way as the one in the Verilog example; it fires if there is an increment when the counter contains a 7. The output reflects the code writer's opinion of the problem:

```
#                       150 data_out: 7
# ** Warning: Oi! Counter Overflow
#    Time: 170 ns   Iteration: 1  Instance: /tb_top/DUT
#                       170 data_out: 0
#                       190 data_out: 1
```

FIGURE 4-8. Finding an error in the VHDL Counter

Creating immediate assertions in your code helps you catch bugs faster by flagging them as soon as they happen. Unfortunately, immediate assertions have a limitation. They check only the current status of data.

Very often our bugs happen because our blocks aren't following the sequence in a protocol properly. While immediate assertions can catch obvious problems such as a `read` and `write` being high at the same time, they cannot catch problems that happen over multiple clock cycles. For that, we'll need multicycle assertions.

4.4 Multicycle Assertions

Multicycle assertions check behaviors that happen over time. For example, say you have a protocol with a request signal (`req`) and an acknowledge signal (`ack`). You may require that the `ack` signal respond to the `req` signal within two to four clock cycles. This is a multicycle assertion.

While it was easy to implement immediate assertions with the `assert` statement, the same is not true of multicycle assertions. These are very difficult to write, because they require the assertion to use state machines and threads to follow complex behaviors.

The industry has responded to this difficulty by creating assertion languages. There are currently two such languages:

- **Property Specification Language (PSL)**—This language is popular with VHDL developers. In fact, new versions of VHDL include PSL as part of the language specification.
- **SystemVerilog Assertions (SVA)**—This language is part of SystemVerilog.

These are powerful languages that allow you to create complex multicycle assertions on a few lines of code. Unfortunately, they are also very complicated languages and have a steep learning curve.

We'll be implementing assertions with a third approach. We'll use an open-source library of assertions called the Open Verification Library (OVL). The next chapter is devoted to the OVL.

4.5 How to Place Assertions

Despite the overwhelming evidence that assertions cut debug time in half, many engineers resist them. A lot of this resistance comes from the idea that assertions are simply mirrors of behavior that is already written into the code. These engineers say, "You're asking me to write code that does something and then write assertions to make sure that the code does what I told it to do. It's a waste of time!"

There are debates about how many assertions we should place inside a block. We are going to sidestep those debates because, in our use-model, we will not be using assertions to check behavior *within* a block; instead we will be using them to check behavior *between* blocks.

Bugs in the communication between blocks can happen easily when there are multiple designers on the same project. They can also happen when we are writing multiple blocks within our own design. We can make mistakes that disrupt communication between our blocks, and these mistakes can be hard to find.

We get maximum value from assertions when we organize them into two categories:

- **Firewall Assertions**—Designers use these assertions to make sure that their blocks are being used properly.
- **Protocol Monitors**—Verification engineers use these assertions to make sure that blocks are communicating properly.

These names reflect the way we use assertions, not how they are implemented. Both of these kinds of assertions can be immediate or multicycle assertions. They can be implemented with the `assert` keyword or the OVL. The key is understanding what they do.

4.5.1 Creating Assertion Blocks

Whether we are creating firewall assertions or protocol monitors, we follow the same strategy for encapsulating assertions. We create a block that takes all the interesting signals as inputs and instantiate it at the top level of our block, or connect it to a communication bus between blocks. Then we write our assertions in this one encapsulating module.

This approach has several advantages:

- It allows firewall assertions to travel with a block and test the inputs.
- In Verilog '95 we can put our assert task into this module and use it for all our assertions.
- The protocol monitors can be reused on future designs if the protocol stays the same.
- When we synthesize assertions in the future, the block can have a single output to show that an assertion fired.

All of our synthesis examples have a single block to hold assertions, whether they are firewall assertions or protocol monitors. Let's look at each of these usage models more closely.

4.5.2 Firewall Assertions

It is Friday afternoon on a clear summer day and you are driving home when your cell phone rings. It is one of the other designers on your project. He's calling to let you know that there is a bug in your block, and that it needs to be fixed by Monday so the team can hit its milestone.

You swear as you realize that your weekend is ruined. Such is the life of an engineer, so you are at your desk Saturday morning fixing the bug.

The debugging is slow, because you can't understand why your block would act this way, but by the afternoon it comes to you. The bug is not in your code at all! Someone else is using your block improperly, and so a bug has propagated into your design.

You call the guy but he doesn't answer his cell phone. He is enjoying the weekend while you're stuck in the office. If only you had installed firewall assertions your roles would be reversed.

Designers install firewall assertions to make sure that bugs outside their blocks stay outside their blocks. These assertions fire when somebody uses your block improperly. They represent your understanding of the specification and how your block should work. They test your assumptions to make sure your block works correctly.

Firewall assertions go into an assertion block at the top level of your design block. They are automatically instantiated when someone uses your design, so you are always checking that your block is being used properly.

For example, if you wrote the three-bit counter and you assumed that nobody would increment your counter if it were full, then you'd test that assumption with an assertion in a firewall module.

Here is the three-bit counter with a firewall assertion module:

```
1  module threebitcounter (input clk, input rst,
2     input ld, input inc,
3     input [2:0] data_in,
4     output reg [2:0] data_out);
5
6     firewall U_check(.clk(clk),.rst(rst),.ld(ld),
7     .inc(inc),.data_in(data_in), .data_out(data_out));
8
9     always @(posedge clk)
10       if (rst)
11          data_out = 0;
12       else
13          if (ld)
14             data_out = data_in;
15          else if (inc)
16             data_out = data_out + 1;
17
18 endmodule // threebitcounter
19
20 module firewall (input clk, input rst, input ld, input inc,
21                  input [2:0] data_in, input [2:0] data_out);
22
23 // synthesis translate_off
24    always @(posedge clk)
25       if (inc && !rst)
26          assert(data_out < 3'h7);
27 // synthesis translate_on
28
29 endmodule
```

FIGURE 4-9. Using a Firewall Assertion Module to Check Input

There are several differences between this block and the one in Figure 4-1 on page 41:

- **Lines 6–7**—We instantiated our firewall at the top level. It takes all our signals as input.
- **Line 16**—The increment block is much simpler, now that we've removed the `assert` statement from the `always` block.
- **Lines 20–29**—We've encapsulated the `assert` statement in its own module.
- **Lines 23 & 27**—We control synthesis from within the firewall module.
- **Line 25**—We qualify our assertion with the `rst` signal, so we don't get spurious assertions.

When we run this code we get the same result as in Figure 4-4 on page 44. The difference is that we've got the assertions encapsulated in one place.

4.5.3 Protocol Monitors

The problem with a blind spot is that you can't see it. For this reason, many FPGA design teams break up the tasks of design and verification. Rather than let the designer create tests that simply validate the designer's potentially flawed assumptions, these teams ask another engineer to look at the block and test it from another point of view.

Some teams have engineers designated only for verification. This ensures that you have a group of people who view it as their mission in life to find bugs in other people's code. This improves quality because, deep down, none of us wants to find bugs in our own code.

Other design teams have engineers take the role of designer on one block and the role of verification on someone else's block. This sort of cross-pollination makes it more likely that both blocks will have higher quality.

In all these cases, the verification engineers have a common constraint. They are not allowed to modify the RTL. In fact, they very often don't know what's in the RTL. They simply have access to the input and output ports. Protocol monitors help these engineers.

Protocol monitors look at blocks and the communication among them to make sure that they are communicating properly. Very often these communications are done with standard protocols such as Wishbone[5] or OCP,[6] but they can also be done with custom communication protocols. In both cases, we need monitors to make sure all the blocks are following the protocol.

If you are using a standard bus such as OCP, you can get protocol monitors from EDA companies. Several are shipped free with simulators such as Questa and Incisive. Others need to be purchased.

In all cases, protocol monitors find difficult bugs quickly because they catch cases where two engineers read a protocol specification and interpreted it differently.

5. The Wishbone bus is an open source bus available on www.opencores.org
6. The Open Core Protocol bus: www.ocpip.org

4.6 Summary

In this chapter we learned about assertions and how to use them. Assertions watch design behavior and fire if they see the behavior deviating from the expected.

There are two kinds of assertions. Immediate assertions check the status of signals at a clock edge and fire if the signals indicate a bad state. Multicycle assertions check the timing relationship between signals and make sure they progress through their timing properly.

Assertions are most valuable when they are checking the communication assumptions between blocks. Designers use firewall assertions to make sure that people follow the designer's assumptions when using a block. Verification engineers use protocol monitors to make sure that the blocks in a design are communicating properly.

Teams should encapsulate their assertions into blocks that sit at either the top of a design block (firewall assertions), or at the top level of a design where the blocks talk to each other (protocol monitors.)

There are assertion languages that allow engineers to create both immediate and multicycle assertions with compact code. The problem is that these languages (PSL and SVA) have a steep learning curve. Instead of trying to learn these languages, we are going to use the Open Verification Library to implement our assertions. This is the subject of the next chapter.

5 The Open Verification Library

- *Introduction to the OVL*
- *What are Checkers?*
- *Downloading the OVL*

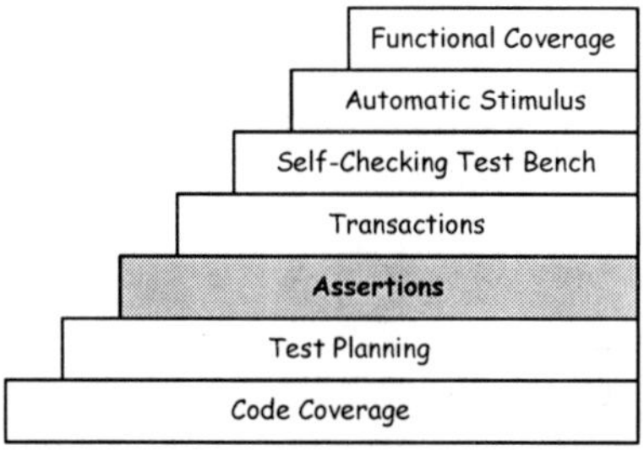

In the preceding chapter, we learned about immediate assertions and how to implement them with the `assert` statement. This is a good start, and if you did nothing but use the `assert` statement effectively you'd be ahead of the majority of design teams in the market.

Immediate assertions are just the start. Multicycle assertions catch the really difficult bugs, but are difficult to write. We need a way to add multicycle assertions to our mix without climbing a steep learning curve. This is why Accellera developed the Open Verification Library (OVL).

Accellera is an industry standards organization created by electronic design and electronic design automation companies. Here is their mission statement:

> *Accellera's mission is to drive worldwide development and use of standards required by systems, semiconductor and design tools companies, which enhance a language-based design automation process.*[1]

The OVL is a library of fifty components that you can instantiate in your design to check its behavior. There are blocks available in SystemVerilog, Verilog95, and VHDL.

1. You can learn more at www.accellera.org.

The blocks in the OVL are called *checkers*, because they do more than implement single assertions. OVL checkers have three advantages over writing assertions ourselves with the `assert` statement:

1. Like all open-source code, the OVL checkers have been tested by real users across the industry. So you don't need to spend time debugging them as you would with your own assertions.
2. OVL checkers implement a set of assertions that completely check a type of functionality, rather than just one assertion to check one set of signals.
3. Ten of the OVL assertions can be synthesized. This means they work in the lab and raise a signal if there is an error deep in your chip. You can attach this signal to a logic analyzer or an LED.

5.1 Checkers vs. Assertions

The OVL is a library of checkers, not assertions. While assertions verify the behavior of a small set of signals in one way, checkers think about things at a higher level. They use a set of assertions to check a behavior completely.

For example, let's consider a design with a set of handshake signals called `req` and `ack`. Let's say that you are creating a protocol monitor to watch the behavior between two blocks as they use `req` and `ack`. You have the following waveform for `req` and `ack`:

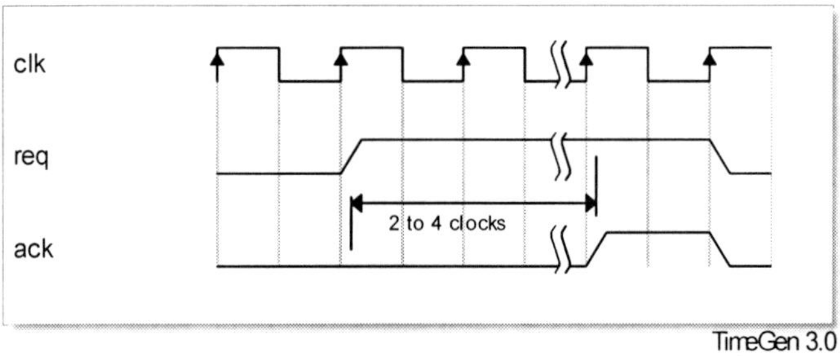

FIGURE 5-1. Example `req/ack` Behavior

The specification says that the ack signal must appear from two to four clock cycles after the req signal and that the ack is one clock wide.

You notice that the OVL has a checker called ovl_handshake. We see here how a checker approaches higher levels of behavior than do simple assertions. A simple assertion might check that the ack signal appears when it should, but there is a lot more to a handshake interface than that. The ovl_handshake checker creates assertions to check the following rules:

- The req signal should not transition again before the ack signal responds.
- The ack signal should not assert if there was no req.
- The ack signal should not go high in fewer than two clocks.
- The ack signal should not take more than four clocks to go high.
- The req signal should not go low before it is acknowledged.
- The req signal should not get stuck high after it is acknowledged.

There is a lot more to a handshake than checking the time from req to ack, and the ovl_handshake checker looks at all of these. You can also program the checker to turn most of these checks off, so if you just want to check the time from req to ack or if your req signal behaves differently, you can avoid spurious errors.

5.2 OVL Languages

EDA is "blessed" with many languages. The OVL has checkers written in four of them.

TABLE 5-1. OVL Assertion Languages

Language	Number of Checkers	Synthesizable Checkers
Verilog '95	33	10
VHDL 93	10	8
SystemVerilog Assertions (SVA)	50	10
Property Specification Language (PSL)	33	0

- **Verilog '95**—This is the granddaddy of all Verilog. It synthesizes in any synthesis tool and simulates in all simulators.

- **VHDL 93**—This language is new to the OVL. It is most useful if you don't have access to a mixed-language simulator. Otherwise you would use the Verilog checkers, because there are more of them.
- **PSL**—Property Specification Language is an assertion language that can describe complex assertions in one line. PSL will be VHDL's native assertion language in the next version of VHDL.
- **SVA**—SystemVerilog Assertions describe the assertion language built into SystemVerilog.

Verilog 95 is the most flexible language in the mix. It works with any Verilog simulator and it has 33 checkers. VHDL93 works with any VHDL simulator. While it has only eight checkers, they are the most popular checkers in the library.

SVA and PSL require advanced simulators. If you have one of these simulators you should use SVA, because it implements 50 checkers.

This means that the only reason you would use the PSL implementation is if you found a simulator that supported PSL but not SVA. I have never heard of such a simulator, so you probably won't be using the PSL checkers.

5.3 Downloading the OVL

The first step to reaping the benefits of the OVL is to go on the web and download the library.[2] The OVL is at www.eda.org/ovl. Once you've accepted the terms and conditions you'll be able to download the gzipped TAR file that contains the OVL. On Windows you can uncompress a file like this with Winzip or any number of freeware decompression tools. On Unix you use `gunzip` and then `tar`.

Unzipping and untarring the file gives you the OVL distribution directory called `std_ovl`. This directory contains unencrypted source code and documentation in HTML and PDF format.

2. OVL Version 2.3 was released in June of 2008. This is the version that we'll be discussing in this chapter.

Here are the contents of the release directory:

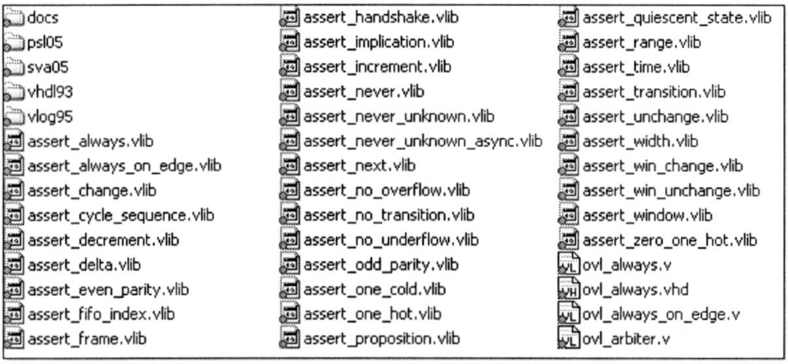

FIGURE 5-2. OVL Distribution Directory

- **docs**—The PDF and HTML documentation for the OVL
- **psl05**—The PSL 2005 version of the OVL. Works with simulators that support PSL but not SVA.
- **sva05**—The SVA 2005 version of the OVL. Works with simulators that support SystemVerilog 3.1a.
- **vhdl93**—The VHDL 93 version of the OVL. Works on VHDL simulators
- **vlog95**—The Verilog 95 version of the OVL. Works with basic Verilog simulators.
- **stdovldefines.h**—A file you should include in your Verilog code if you want to use the OVL constants.
- ***.vlib**—Legacy source code for the OVL. This is the source code from OVL 1.0, but all these assertions have been deprecated because they are not synthesizable.
- ***.v**—The source code for the OVL. All synthesizable OVL assertions start with the string `ovl_`.

This is the root directory for the OVL. Whenever we compile the OVL or reference it in a compilation script, we use this directory on our command line. Now that we have the OVL downloaded, we can use it in our first synthesizable assertion. Because the use models for Verilog and VHDL are different, each gets its own chapter. Let's look at Verilog, the subject of our next chapter.

6 Verilog Library Primer

- *Verilog Modules*
- *Port Mapping in Verilog*
- *Verilog Parameters*
- *Verilog* `include` *Files*
- *Verilog Macros*
- *Conditional Compilation*

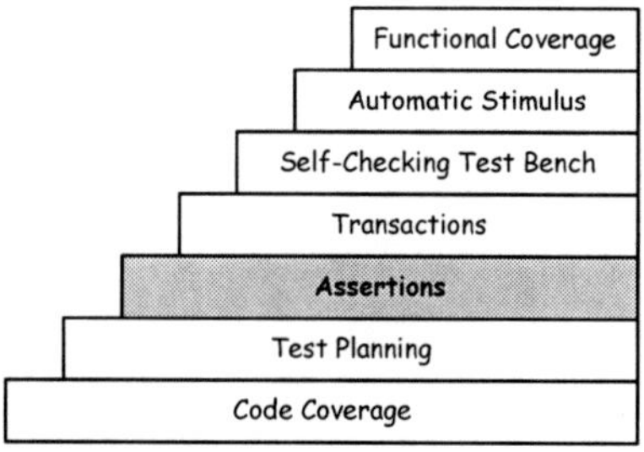

This chapter is for VHDL users who are not familiar with Verilog library management. Experienced Verilog users can safely skip this chapter. The SystemVerilog and Verilog OVL chapters assume the reader understands the material in this chapter.

Verilog library management is philosophically based on C.[1] There are two basic differences between Verilog and VHDL library management:

- **Verilog Uses a Preprocessor**—This is an artifact of Verilog's C heritage. It uses a preprocessor to massage source files before it begins compiling them. This means that it controls libraries by choosing what code to insert or compile and what code to ignore. VHDL compiles everything and lets the configurations figure it out.
- **Verilog Uses the File System for Library Management**—In another homage to C, Verilog makes assumptions about how you'll name your source files and asks you what directories it should look in. In VHDL, once everything is compiled, you just use the library names to find components.

All the library management tools in Verilog help us define which modules get compiled and where to find them. So we'll start our Verilog library discussion with modules.

1. As opposed to VHDL, which is philosophically based on Ada.

6.1 Verilog Modules

Where VHDL has components, Verilog has modules. The big difference between them is that Verilog has no concept of an entity/architecture pair. VHDL separates the definition of the interface to a component (the entity) from the behavioral description (the architecture.) Verilog combines both of these in the module definition.

This raises the question, "What if I want to have one module with two representations—such as RTL vs. gate?" In that case you define two modules with the same name and compile only one of them. That is the basis of Verilog library management.

6.2 Port Mapping in Verilog

When you instantiate a module in Verilog or SystemVerilog, you can explicitly name the ports connected to each signal. This is the recommended way of connecting signals to modules. Here's an example of creating two modules and mapping signals to ports:

```
module top (output out, input a, b);

    rayand U_ray (.prod(out), .in1(a), .in2(b));

endmodule

module rayand(output prod, input in1, in2);

   assign prod = in1 & in2;

endmodule
```

FIGURE 6-1. Verilog Port Mapping

We'll be using this port mapping feature extensively when we instantiate OVL modules, because it is unlikely that the OVL port names will match our design's signal names.

6.3 Verilog Parameters

VHDL and Verilog both have a mechanism that allows you to change constant values at instantiation. In VHDL these values are called *generics*. In Verilog these values are called *parameters*.

Parameters set values that cannot change during the simulation, such as bus widths. Parameters control an OVL checker's behavior. For example, there is a checker called `ovl_width` that makes sure a signal is asserted for a minimum and maximum number of clock cycles. You set the checker's minimum and maximum clock cycles with a parameter.

Here is a parameter example that uses parameters to change the behavior of a module instantiated several times:

```
 1 module petlove;
 2    parameter pet = "animals";
 3    parameter color = "all";
 4    initial
 5       $display("I love %s %s",color, pet);
 6 endmodule
 7
 8 module top;
 9    petlove #(.pet("dogs"))                   U_dogs();
10    petlove #(.color("brown"))                U_brown();
11    petlove #(.pet("cats"),.color("black")) U_cats();
12 endmodule
```

FIGURE 6-2. Parameters Controlling Constants

The `petlove` module has two parameters: `pet` and `color`. The module designer defines them with the `parameter` statements on lines 2 and 3.

The instantiation sets the parameter with the `#(...)` syntax. Notice that the parameter definitions go before the instance name. We connect to the parameter with the same syntax as the port list.

When we run this example, we see that our parameters were passed correctly:

```
# Loading work.top
# Loading work.petlove
# I love all dogs
# I love brown animals
# I love black cats
```

FIGURE 6-3. Parameters in Output

The OVL LRM defines the parameters for each checker.

6.4 The `-y` and `+libext+` Command Line Options

VHDL has the concept of a library, which is a precompiled set of components that you can reference from your source code with the `LIBRARY` statement. This is quite handy, because it makes library references explicit to people reading the code.

Verilog does not have this concept. There is nothing in the source code that points to an explicit library or matches an entity and architecture. Instead, Verilog takes a module name and searches a list of directories for a file whose name matches the module name with an extension. When it finds the file, it compiles it.

This means that the same Verilog source code can behave differently depending upon the compiler command line. Many times you will see that Verilog teams have directories called "RTL" and "gates" in their design tree.

When the design team wants to run an RTL simulation, they tell the compiler to look for the modules in the "RTL" directory. When they want to do gate-level simulation, they tell the compiler to look in the "gates" directory. Both directories contain module definitions that have the same names and different implementation, but only one implementation gets compiled.

Verilog compilers assume that modules are stored in a file with the same name as the module plus an extension. It is up to the person invoking the compiler to tell it what extension should be added to module names and where it can find the files with those file names.

For example, let's say I've created a library element called rayand (my own AND gate) and stored it in a directory called raylib. Here's how I'd use that library element:

```
module top (output out, input a, b);

    rayand U_ray (.out(out), .a(a), .b(b));

endmodule
```

top.v

```
module rayand(output out, input a, b);

   assign out = a & b;

endmodule
```

raylib/rayand.v

FIGURE 6-4. A Library Example

The top module in this design instantiates rayand. The rayand module is not defined in top.v or any other file on the command line, so Verilog will go looking for a file with the name rayand plus an extension. We will tell the compiler how to find the file with the -y and +libext options.

Here are their definitions:

- -y <directory name>—This stands for librar**Y** (or, I suppose, director**Y.**) It tells the Verilog compiler which directories contain library files. If you have multiple -y command line options, they form a search path based on their order on the command line.
- +libext+<extension>—This is the extension that goes on all library file names. When the Verilog compiler cannot find a module, it takes the module name and appends the library extension to it to make a file name. Then it looks in the library directories to find the file.

In complex systems there may be many -y and +libext options on the command line. Design teams normally store complex command line options like this in a file. (See "The -f Command Line Option" on page 71.)

We use these command options to tell the Verilog compiler how to find the definition of `rayand`:

```
vlog +libext+.v -y raylib top.v
```

The `+libext+.v` option told Verilog to take the module name (`rayand`) and append the string "`.v`" to the end to make `rayand.v`. The `-y` option told Verilog to look for `rayand.v` in the `raylib` directory.

When we compile in ModelSim we can see Verilog looking for the module:

```
# -- Compiling module top
# -- Scanning library directory 'raylib'
# -- Compiling module rayand
#
# Top level modules:
#       top
```

FIGURE 6-5. Searching for a Library Element

We use this approach to instantiate OVL modules in Verilog and SystemVerilog. The only difference is that the `-y` option points to the `std_ovl` release directory.

6.5 Verilog `include` Files

A preprocessor is a program that modifies your source code before it goes into the compiler. The preprocessor can remove whole chunks of code, change the value of strings, and include text from a file before compiling.

In this section we'll look at including text with the `` `include `` preprocessor command. All Verilog preprocessor commands start with the backtick (`` ` ``) character, so we include a file in our code by typing

```
`include "filename"
```

The `` `include `` acts exactly as if we had opened the included file in a text editor, copied the text, and pasted it into our source file in place of the `` `include `` line. A file that's been included this way is called an *include file*.

The compiler assumes that include files are in the same directory as the source file. This is a trivial case. In reality, you want to be able to store your include files in a separate directory. We use the `+incdir+` option to tell the Verilog compiler where to find include files.

Let's solve our `rayand` library problem by using an include file:

```
1  `include "rayand.h"
2
3  module top (output out, input a, b);
4
5      rayand U_ray (.out(out), .a(a), .b(b));
6
7  endmodule
```

top.v

```
module rayand(output out, input a, b);

   assign out = a & b;

endmodule
```

raylib/rayand.h

FIGURE 6-6. Using `'include`

On line 1 we've used `'include` to add the definition of the `rayand` module to the `top.v` file. The preprocessor inserts the file `rayand.h` into `top.v` and then compiles it.

The `rayand` module is stored in a file called `rayand.h`. The `h` extension comes from the word "header" and is a C convention that's been brought over to Verilog (some people use `.vh`).

We compile this design with the following command line using `+incdir`:

```
vlog +incdir+raylib top.v
```

We could have specified multiple directories with `+incdir` to create a search path.

6.6 Verilog Macros

Verilog macros use the preprocessor to replace a string in the source code with another string. Macros allow us to replace constants in our source code (say, the number `32`) with meaningful words (such as `BUSWIDTH`). This makes the code readable and allows us to make wholesale changes to a design by changing only one line.

Like all preprocessor commands, macro definitions and macro usages start with the backtick[2] (`` ` ``) character. We define macros with `` `define `` like this:

`` `define *MACRONAME* [*macrovalue*] ``

The macro name does not need to be in all uppercase, but this is a convention used across the industry. The macro value is optional; you can define a macro without giving it a value. This is useful when we talk about conditional compilation.

We reference macros with the backtick and the macro name, so a macro reference looks like this:

`` `*MACRONAME* ``

When the preprocessor sees this, it replaces the macro reference with the string that was defined earlier. This has the same effect as doing a global search and replace in an editor.

2. Under the tilde (~) on the top left of most keyboards.

Here is our `petlove` example with macros:

```
`define BIGPET1 "dogs"
`define FAVCOLOR "brown"

module top;
   petlove #(.pet(`BIGPET1))        U_dogs();
   petlove #(.color(`FAVCOLOR))     U_brown();
endmodule


module petlove;
   parameter pet = "animals";
   parameter color = "all";
   initial
      $display("I love %s %s",color, pet);
endmodule
```

FIGURE 6-7. Macros in Action

Developers rarely define their macros in the same file where they are used. Usually the macros get defined in a header file and then get used in compilation. The preprocessor executes all the `` `include `` commands first, and then it replaces the macros.

The `petlove` example looks like this with an included header file:

```
`include "petinfo.h"

module top;
   petlove #(.pet(`BIGPET1))        U_dogs();
   petlove #(.color(`FAVCOLOR))     U_brown();
endmodule


module petlove;
   parameter pet = "animals";
   parameter color = "all";
   initial
      $display("I love %s %s",color, pet);
endmodule
```

`top.v`

```
`define BIGPET1  "pigs"
`define FAVCOLOR "pink"
```

`petinfo.h`

FIGURE 6-8. Using a Header File with Macros

This approach of using the header file to define the macros allows us to control module behavior without changing our Verilog code. We can control the `pet-love` modules simply by changing the values in `petinfo.h`.

We'll see that using the OVL requires us to include the `std_ovl_defines.h` at the top of files that instantiate OVL elements. This is where the OVL defines all its macros.

6.7 Conditional Compilation

Verilog's preprocessor allows us to conditionally control whether the compiler sees a block of code. If we use conditional compilation to disable a block of code, then it is as if we had used a text editor to remove the code before we compiled.

In the last section we said that you could define a macro without giving it a value. You would do this to control conditional compilation. The conditional compilation commands look at whether a macro has been defined to make decisions.

There are five commands that control conditional compilation:

- `'ifdef <MACRO>`—This command compiles the code in its block if the macro is defined.
- `'ifndef <MACRO>`—This command compiles the code in its block if the macro is *not* defined.
- `'elsif <MACRO>`—This command follows an `'ifdef` or `'ifndef`. It checks for its macro if the previous command was false.
- `'else`—This command follows any of the conditional macros above. It compiles its code if the macro above is false.
- `'endif`—This command closes the conditional compilation block.

The conditional compilation in Verilog is much more drastic than the conditional compilation in VHDL. VHDL's conditional compilation compiles all the code but then makes elaboration time decisions as to whether to use it. It does this with the `generate` statement.

Verilog conditional compilation hides the code, so that the compiler never sees it. This is useful for cases where the code would cause a syntax error, perhaps because of language version issues.

As an example, let's go back to our three-bit counter and the assertion module that we placed in it. We are going to use conditional compilation to accomplish two things:

1. We will allow the user to remove the assertion module completely.
2. We will modify the assertion module so that it can be compiled with either a Verilog compiler or a SystemVerilog compiler.

We will use conditional compilation and macro definitions to implement these requirements. We'll give the module user two macros that control compilation:

- `ASSERTIONS_OFF`—This macro removes all the assertions from the design.
- `VERILOG`–This macro allows the files to be compiled with a Verilog '95 compiler

Once again we see the wisdom of encapsulating all the assertions in one module. This allows us to place all the assertion controls in one place. The person instantiating the `threebitcounter` doesn't need to know about any of this code. That person just needs to define the macros.

Here's the assertion module that implements these requirements:

```
module threebitcounter_firewall(
      input clk, input rst, input ld, input inc, input [2:0] data_in,
      input [2:0] data_out);

   // synthesis translate_off

`ifndef ASSERTIONS_OFF

`ifdef VERILOG
     task assert;
       input condition;
       begin
          if (!condition)
             $display("Assertion Error");
       end
     endtask
`endif

always @(posedge clk)
   if (!rst && !ld && inc)
     assert(data_out < 3'h7);

`endif

  // synthesis translate_on
endmodule
```

FIGURE 6-9. Assertion Module with Conditional Compilation

- **Line 7**—This line simulates the assertions unless the user defines the `ASSERTIONS_OFF` macro.
- **Line 9**—We can nest conditional code. This code is compiled only if the user defines the `VERILOG` macro. This means that we could compile this code in a SystemVerilog compiler, because the compiler would never see the task definition on lines 10–16.
- **Line 17**—The `` `endif`` matches the last `` `ifdef``. This closes the block that started on line 9.
- **Line 23**—This closes the block that started on line 7.

If the user turns off assertions with the `ASSERTIONS_OFF` macro, we will still compile the `threebitcounter_firewall` module. It's just that it will be empty.

While we defined macros earlier with lines in the source code, Verilog compilers allow you to define macros on the command line. So rather than put a `` `define`` into our source code, we can turn the assertions off by defining `ASSERTIONS_OFF` on our command line:

```
vlog threebitcounter.v +define+ASSERTIONS_OFF ...
```

Using the `+define` command line option is the same as placing a `` `define`` at the top of every file. When we compile this code we will get the three-bit counter without its assertions.

We compile the `VERILOG` conditional code like this:

```
vlog threebitcounter.v +define+VERILOG ...
```

This option defines the Verilog macro and thus compiles our `assert` task.

You can define multiple macros on the same line:

```
vlog threebitcounter.v +define+
   ASSERTIONS_OFF+VERILOG ...
```

However, you need to be careful about two things when you use macro definitions:

1. Macro names are case sensitive. If the conditional compilation lines are looking for macros with the names in all capitals, you need to put them in all capitals on the command line.

2. There is *no* spell checking on the macro names. If you type `+define+ASSERTION_OFF` (as opposed to `ASSERTIONS_OFF`, missing the third 'S'), then your assertions will be compiled, because the Verilog compiler will define a macro called `ASSERTION_OFF` that the design never uses.

The OVL uses macros and conditional compilation everywhere. For example, there is a macro called `OVL_ASSERT_ON`. You need to define this macro if you want the checkers to contain assertions.

6.8 The -f Command Line Option

The Verilog compiler command line can get very long in a complex environment. There are macros to define, libraries and include directories to reference, and then all the files that need to be compiled.

Verilog compilers address this problem with the `-f` command line option. This option replaces the command line with a file that contains all the information that would have gone on the command line.

Here is an example of using the `-f` option to shorten a script. This script runs in ModelSim, though similar scripts run in all simulators. It is called `run.do` and it will be the basis for all the example scripts in the design.

```
1  if [file exists work] {vdel -all}
2  vlib work
3  vlog -f compile.f
4  vsim top;
5  run -all
```

FIGURE 6-10. `vlog` Command using `-f`

- **Line 1**—Remove the existing simulation library if it exists.
- **Line 2**—Now make a new library.
- **Line 3**—Compile all the Verilog files. This is a very simple command in this script because we've hidden all the command line options and file names in the file called `compile.f`.
- **Line 4**—Launch the simulation.
- **Line 5**—Run the simulation to the end.

This script looks short and simple, because I've put all the arguments to the command in a file called `compile.f`. By convention I use the `.f` extension on these files. This matches the `-f` option. Here is the `compile.f` file that compiles the three-bit counter with a Verilog assertion:

```
1 top.v
2 threebitcounter.v
3 -y . +libext+.v
4 -vlog95compat
5 +define+VERILOG
```

FIGURE 6-11. `compile.f` **File for** `threebitcounter`

- **Lines 1 & 2**—We list the two source files, but we don't list the file containing the firewall assertions. We'll search for that by using the library options.
- **Line 3**—We use the `-y` and `+libext` options to tell the Verilog simulator where to find library modules. We want Verilog to find the `threebitcounter_firewall` module that was defined in Figure 6-9 on page 69.
- **Line 4**—The `-vlog95compat` option forces Questa to compile with Verilog syntax instead of SystemVerilog. We need this because we created a task called `assert`, and the word `assert` is a keyword in SystemVerilog.
- **Line 5**—I define the VERILOG macro so that the `assert` task will be compiled. If we were compiling with SystemVerilog, we would not define this macro. We would use the built-in `assert` statement instead.

When we use the OVL we will always use a file like this, because there are several options that we always need to set to make the OVL work.

6.9 Summary

This chapter was a primer in Verilog library management for readers who have little experience with Verilog. These concepts are key to using the Verilog and SystemVerilog OVL.

We covered modules and discussed their differences from VHDL components. The key to Verilog library management is to see that Verilog uses the file system as a library management tool. Instead of compiling all the possible components and using configuration files to manage them, Verilog relies upon command line options and search paths to compile only the files that are needed.

Once those files are chosen, we control how they are compiled by using macros. We either replace strings in our source code with macro strings, or we use the macros to control conditional compilation.

Finally, we store all these options in `.f` files to make our scripts easier to read. In the next chapter we'll start using the Verilog and SystemVerilog checkers in the OVL.

7 Using the OVL with Verilog and SVA

IN THIS CHAPTER:

- *The Assertion Module*
- *Instantiating OVL Modules*
- *Compiling the OVL*

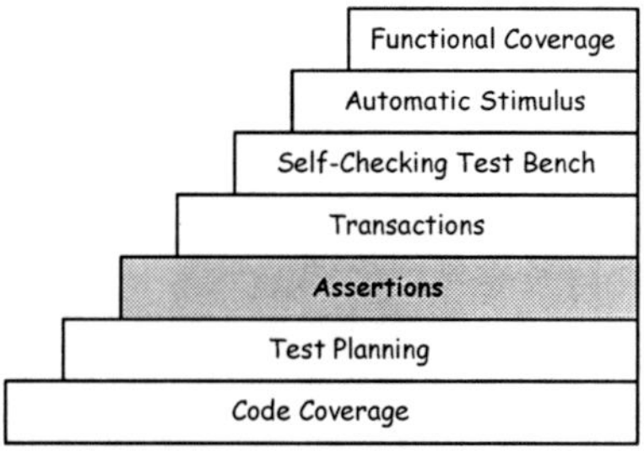

The Open Verification Library (OVL) has checkers written in Verilog, SystemVerilog Assertions (SVA), and VHDL.[1] Because the checkers are written for us, we don't need to know these languages in order to take advantage of them. Instead, we'll just instantiate the modules and let the checkers do the work.

The most advanced checkers are the SVA checkers, followed by the Verilog checkers, and then the VHDL checkers. We use the SVA checkers if we have a simulator that works with SystemVerilog Assertions. Because Verilog and SVA work the same way, we'll discuss them together in this chapter.

This chapter assumes that you are familiar with the topics we discussed in Chapter 6, including Verilog library management with module instantiation, parameters, and the `-y` option.

---

1. There are also PSL checkers but we don't use them since the SVA assertions provide more functionality.

7.1 The Assertion Module

In our previous chapters we discussed encapsulating all our assertions in an assertion module and then instantiating this module at the top of a block as a fire-wall or between blocks as a protocol monitor.

This approach has two benefits. First, it makes it easy for someone reusing the code to find all our checkers. Second, it makes it easy to use Verilog and SVA checkers in a VHDL design.

For example, here is our VHDL three-bit counter instantiating a Verilog assertion module:

```
BEGIN // ARCHITECTURE rtl of threebitcounter

   U_firewall : assertion
      PORT MAP (clk => clk, rst => rst,
                ld => ld, inc => inc,
                data_in => data_in, data_out => data_out);

   main_proc : PROCESS (clk)
      BEGIN
         IF (clk'EVENT AND clk = '1') THEN
            IF (rst = '1') THEN
               data_out <= "000" ;
            ELSE
               IF ld = '1' THEN
                  data_out <= data_in;
               ELSIF inc = '1' THEN
                  data_out <= std_logic_vector(data_out + "001");
               END IF;
            END IF;
         END IF;
   END PROCESS main_proc;
END rtl;
```

FIGURE 7-1. Verilog Assertion Module in VHDL Three-Bit Counter

The `assertion` module contains all our Verilog or SVA checkers. As we'll see below, we need to pass parameters to these checkers and map ports to them. This could be tricky if we were instantiating the checkers directly in a VHDL environment. By using the `assertion` module, however, we sidestep all these issues and instantiate a simple module with a simple port map.

We know how to set up the libraries to support a Verilog assertion module, as we discussed in Chapter 6. Now we just need to see how to find checkers in the OVL documentation and translate them into code.

7.2 Instantiating OVL Modules

Now that we've created an assertion module to watch our design, we need to instantiate the OVL checkers in that module. There are three steps we need to follow for each checker:

1. Decide what needs to be checked.
2. Find a corresponding OVL checker using the OVL LRM.
3. Figure out the correct parameters and instantiate the checker.

We'll follow these steps with two examples. One example is our friend the three-bit counter. This is an immediate assertion implemented with the OVL.[2] The second example is a request-and-acknowledge handshake. We'll watch that one with a multicycle assertion.

7.2.1 Testing the Three- Bit Counter with the OVL

In this section, we'll create a firewall assertion for our three-bit counter. The firewall assertion checks to see that we never increment a full counter. We never want to see an increment signal be high when the counter contains the number `3'h7`.

This accomplishes step 1: we've decided what needs to be checked. Now we need to find an OVL checker to do the job. The OVL ships with a file called `ovl_quick_ref.pdf`. This is a single sheet that summarizes all fifty checkers. Scrolling part way through this sheet we find the following:

Single-Cycle	**ovl_never**	test_expr must never hold

FIGURE 7-2. The `ovl_never` Entry in the Reference Sheet

The Description column for this entry says, "`test_expr` must never hold," which means that `test_expr` must never be true. This is exactly what we need.

2. Why would we bother using the OVL for an immediate assertion? Check out Chapter 7 on synthesized assertions.

Our next step is to look up `ovl_never` in the file `ovm_lrm.pdf`:

ovl_never

Ensures that the value of an expression is not TRUE.

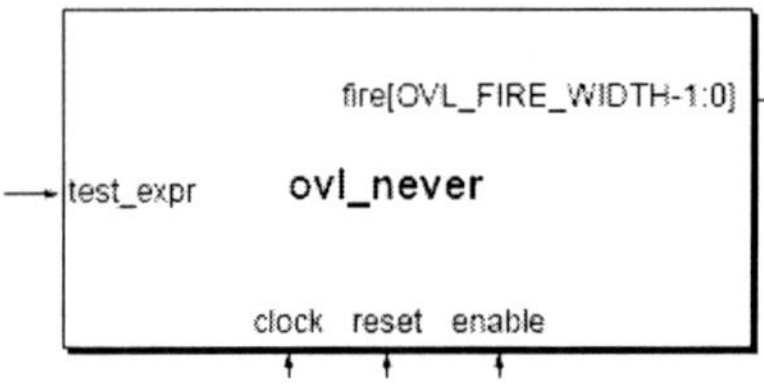

Parameters/Generics:
severity_level
property_type
msg
coverage_level
clock_edge
reset_polarity
gating_type

Class: 1-cycle assertion

Syntax

```
ovl_never
    [#(severity_level, property_type, msg, coverage_level, clock_edge,
       reset_polarity, gating_type )]
  instance_name (clock, reset, enable, test_expr, fire);
```

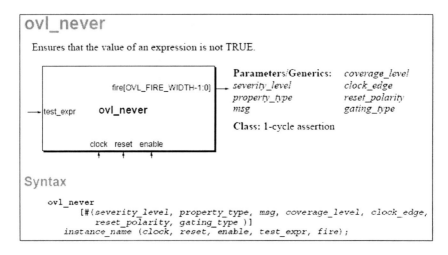

FIGURE 7-3. The `ovl_never` Entry in the OVL LRM[3]

This picture shows us everything we need to instantiate the `ovl_never` checker. The rectangle shows the inputs and outputs from the checker. The inputs on the bottom are common to all checkers. The inputs on the left side are used by this checker. The output `fire` on the right side is common to all checkers.

There are three input ports common to all checkers:

- `clock`—Most checkers work on a clock edge. By default the checker operates on the positive edge of the clock. We'll see how to switch this when we discuss parameters.
- `reset`—This resets the checker back to its original state. It doesn't do anything in a single-cycle checker such as `ovl_never`, but some checkers need to be reset.
- `enable`—You can avoid false error messages by controlling when the checker operates. By default a disabled checker ignores the cycle in which it is disabled. You can change this behavior with a parameter.

3. All references to the Open Verification Library documentation have been copied with the permission of Accellera.

There is one output port that is common to all checkers:

- three-bitfire ['OVL_FIRE_WIDTH-1:0]—The fire bus is several bits wide.[4] We are going to use fire[0]. This signal goes positive when the assertion fires and can be used in a waveform to find assertion errors.

There are seven parameters common to all checkers. The ovl_never checker is so simple that it uses only the seven common parameters. Complicated checkers have parameters unique to them. For example, ovl_width has parameters that allow you to define the minimum and maximum number of cycles a test_expr can remain true.

Here is an overview of the seven common parameters:[5]

- severity_level—You can control how the simulator responds to the assertion. If you set this parameter to 'OVL_FATAL, the simulator stops. By default, the severity is 'OVL_ERROR, which logs an error and print a message.
- property_type—We ignore this parameter.
- msg—We can pass a string that the checker uses instead of the default OVL error message. The string can contain information useful to filtering the log file.
- coverage_level—We discuss this parameter in Chapter 27. We can ignore it now.
- clock_edge—By default, the checkers work on the positive edge of the clock, but we can control that by setting this parameter to 'OVL_POSEDGE or 'OVL_NEGEDGE.
- reset_polarity—By default, the reset signal is active low. We can specify this behavior explicitly by setting this parameter to 'OVL_ACTIVE_LOW or 'OVL_ACTIVE_HIGH.
- gating_type—We control the enable port's behavior with this parameter. By default, the checker ignores clock cycles where the enable signal is low.

  We can specify this behavior by setting this parameter to one of three macros: 'OVL_GATE_NONE, 'OVL_GATE_CLOCK, or 'OVL_GATE_RESET. The first disables the enable. The second is the default. The third causes the checker to reset when the enable is activated.

Once we understand the checker, we simply instantiate it in our assertion module. The -y and +libext options take care of finding the checker code and compiling it (more on those in Section 7.3.)

4. The fire bus is three bits wide in version 2.3. That could change in a future version.
5. The details on these parameters are in the OVL LRM.

Here is the three-bit counter firewall assertion module that was instantiated into the VHDL in Figure 7-1 on page 76. We've connected the counter signals to it and set the necessary parameters.

```
1  `include "std_ovl_defines.h"
2  module assertion (input clk, input rst, input ld, input inc,
3                    input [2:0] data_in, input [2:0] data_out);
4
5  wire [`OVL_FIRE_WIDTH-1:0] fire;
6
7  ovl_never
8
9  #(.reset_polarity(`OVL_ACTIVE_HIGH), .msg ("Counter Overflow"))
10
11 U_counter_overflow
12
13          // Common ports
14          (.clock(clk), .reset(rst), .enable(~rst), .fire(fire),
15
16          // ovl_never port
17          .test_expr((data_out == 3'h7) && inc));
18
19 endmodule
```

FIGURE 7-4. Three-Bit Counter Firewall with OVL

- **Line 1**—All assertion module source files include `std_ovl_defines.h`.
- **Line 5**—We create this internal bus for the fire port. We could have left the port unconnected, but then we'd get warning messages in the simulator.
- **Line 7**—Here's where we instantiate the checker. The name is the same as that in the documentation box.
- **Line 9**—You set parameters in Verilog before you provide the instance name. We are setting two parameters. We have an active high reset and we want to have a message that contains the string `Counter Overflow`.
- **Line 11**—This instance has the name `U_counter_overflow`.
- **Line 14**—These are the ports common to all checkers. Notice that the signal names are different from the port names. This is an example of Verilog port mapping.
- **Line 17**—Verilog allows you to put an expression into a port. Verilog analyzes the expression, and the result goes into the module as an input signal (there is no need to cast the result). In this case we fire if `data_out` is `3'h7` and the `inc` signal is high.

When we run the three-bit counter test the output looks like this:

```
# Loading work.ovl_never
#                         0 data_out: x
#                        10 data_out: 0
#                        30 data_out: 1
#                        50 data_out: 2
#                        70 data_out: 3
#                        90 data_out: 4
#                       110 data_out: 5
#                       130 data_out: 6
#                       150 data_out: 7
#         OVL_ERROR : OVL_NEVER : Counter Overflow : Test expression is not
FALSE : severity 1 : time 170 :
top.DUT.u_firewall.U_counter_overflow.ovl_error_t
#                       170 data_out: 0
#                       190 data_out: 1
#                       210 data_out: 2
```

FIGURE 7-5. Counter Overflow Captured by OVL

The OVL checker caught the fact that we tried to increment a full counter. There are more details here than in our `assert` message. There are several fields separated by colons (:).

- We've got an `OVL_ERROR`. This is the default severity.
- We see it's an `OVL_NEVER` checker. We can use these messages to filter our output file.
- The next string is our message `Counter Overflow.`
- The OVL complains that the test expression is not false.
- The severity level is 1, which is the number assigned to the macro `` `OVL_ERROR``. You can see all these assignments in `std_ovl_defines.h`.
- The error happened at time 170.

The OVL checker has caught our assertion and printed to the screen. That's not all, because in this example we used the SystemVerilog assertion version of the OVL. The SVA version of the OVL uses the SystemVerilog assertion language to define the checker. This allows simulators that support SVA to provide more-detailed debug information when an assertion fires.

In the case of Questa, it not only prints the SVA errors to the screen, but it also makes a note of them in the waveform viewer. Questa can only do this if we use the SystemVerilog assertion version of the OVL.

In this example we've added our assertion to the waveform viewer:

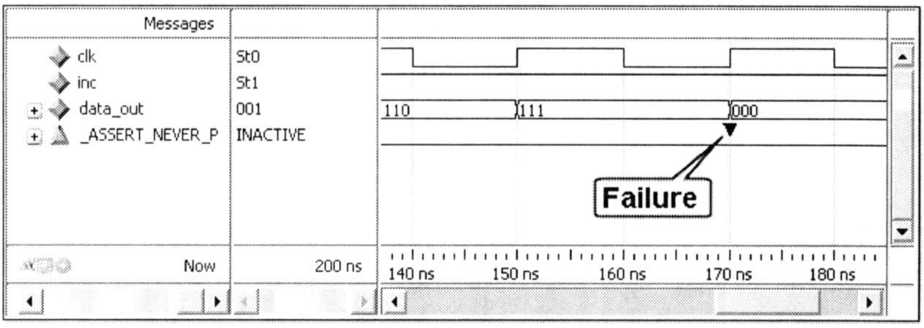

FIGURE 7-6. Viewing an Assertion in a Questa Waveform

You can see the error fire at 170 ns. Questa also creates reports to show which assertions failed in the simulation. We'll see how to designate the SVA version of the OVL in Section 7.3.

7.2.2 Using a Multicycle Checker

In Chapter 4, we discussed the two kinds of assertions: immediate and multicycle. Immediate assertions are like the one we created to check our counter. It can look at values on each clock edge and decide whether to fire. However, the single-cycle assertion is only a special case.

The *OVL Language Reference Manual* defines five different kinds of assertions, depending upon the number of clock cycles it takes for them to operate:

- **Combinational Checkers**—These don't use a clock to fire, but rather check some variable or condition continuously. Checkers that look for the existence of X on a bus fall into this category.
- **1-Cycle Checkers**—These are like `ovl_never`. They check a value every clock cycle.
- **2-Cycle Checkers**—These compare the value on the next clock cycle to the value during the current clock cycle.
- ***n*-Cycle Checkers**—These look at values over a number of clocks supplied by the user.
- **Event-Bounded Checkers**—These checkers use two events to control what they check. For example, `ovl_win_unchange` says that the test expression must not

> change between two events. You may not want the value of a RAM address to change between the time the write signal goes low and then goes high again.

When we instantiate a multicycle checker, we need to provide information to tell the checker how to determine its window of operation. We do this by using a combination of signals and parameters.

The `ovl_handshake` checker is a good example. This checker makes sure that if we see one signal in a protocol, then it is followed by another signal in a window of time. Here's an example:

If the `req` signal goes high, then the `ack` signal must go high two to four clock cycles later.

This is a complicated assertion to write in VHDL or Verilog. It requires that we create a little state machine and then fork off multiple processes running that state machine. Fortunately, it is easy to implement with the OVL.

We use the OVL by searching through the quick reference guide for a checker that might suit our needs. We find the following *n*-cycle checker:

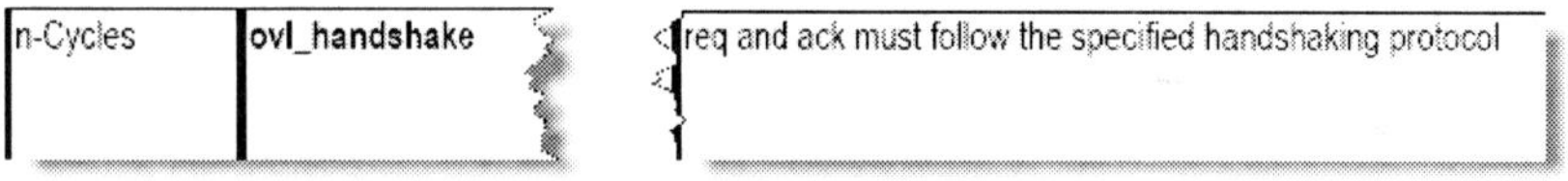

FIGURE 7-7. Quick Reference Entry for `ovl_handshake`

This checker is useful not only for situations with a specific request and acknowledge. It can be used in any situation where we want to make sure that some event eventually follows another event.

When we open `ovl_lrm.pdf` we find the following entry for `ovl_handshake`:

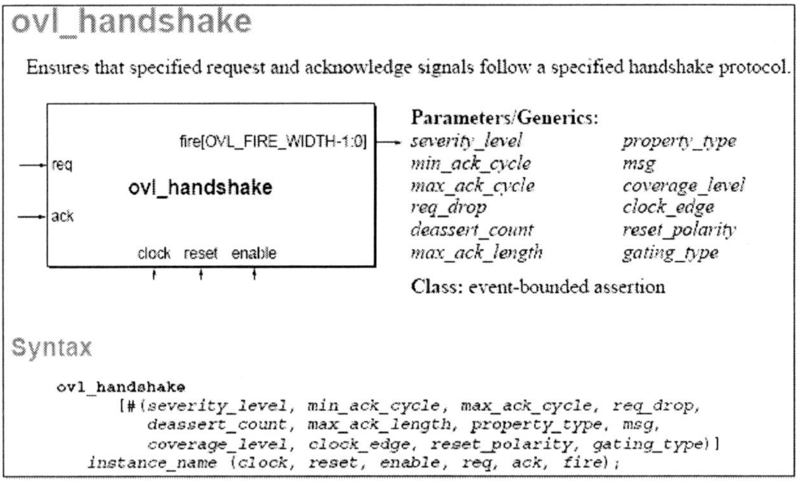

ovl_handshake

Ensures that specified request and acknowledge signals follow a specified handshake protocol.

fire[OVL_FIRE_WIDTH-1:0]
req
ovl_handshake
ack
clock reset enable

Parameters/Generics:

severity_level	*property_type*
min_ack_cycle	*msg*
max_ack_cycle	*coverage_level*
req_drop	*clock_edge*
deassert_count	*reset_polarity*
max_ack_length	*gating_type*

Class: event-bounded assertion

Syntax

```
ovl_handshake
     [#(severity_level, min_ack_cycle, max_ack_cycle, req_drop,
        deassert_count, max_ack_length, property_type, msg,
        coverage_level, clock_edge, reset_polarity, gating_type)]
  instance_name (clock, reset, enable, req, ack, fire);
```

FIGURE 7-8. LRM Entry for OVL Handshake

The LRM says that two input ports are unique to this checker: the request signal (`req`) and the acknowledge signal (`ack`). There are also several parameters unique to `ovl_handshake`. We are interested only in two of them:

- `min_ack_cycle`—This is the minimum number of cycles that must pass before the acknowledgement. In our case this number is two.
- `max_ack_cycle`—This is the maximum number of cycles that can pass before the acknowledgement. In our case this number is four.

Below the box and the list of parameters, we see how to instantiate the checker. We instantiate `ovl_handshake`, supply the correct parameters, and then connect the `req` and `ack` signal.

In our example, we use `ovl_handshake` in a protocol monitor. This is a module that sits at the top level of a design and watches the signals between the blocks. It makes sure that the blocks are obeying the protocol.

Here is how we instantiate the `ovl_handshake` checker:

```
1  `include "std_ovl_defines.h"
2
3  module protocol_monitor(
4     input clk,
5     input reset_n,
6     input req,
7     input ack);
8
9    wire [`OVL_FIRE_WIDTH-1:0] fire;
10
11   ovl_handshake
12     #(.min_ack_cycle (2), .max_ack_cycle(4))
13     check_ack
14     (
15        // Unique ports on ovl_handshake
16      .req(req), .ack(ack),
17
18        // Common ports on all OVL
19      .clock(clk), .reset(reset_n), .enable(1'b1), .fire(fire));
20
21  endmodule
```

FIGURE 7-9. Instantiating `ovl_handshake`

To make things clear, we've broken the instantiation into multiple lines:

- **Line 11**—The name of the checker module.
- **Line 12**—The parameters that tell the checker that we need a minimum of two clocks cycles and a maximum of four.
- **Line 16**—The request and acknowledge signals.

The rest of the signals in the instantiation are the same as in all other OVL checkers.

The `ovl_handshake` checker is a good example of the difference between a single assertion and a complete checker. There is more to checking a handshake than the minimum and maximum time from the request to the acknowledge. We also need to check that the acknowledge happened in response to a request, and that the request waited for the acknowledge before launching another request.

This means that there are several assertions in a single check. The documentation in `ovl_lrm.pdf` shows us all of these:

ovl_handshake

Assertion Checks

MULTIPLE_REQ_VIOLATION	The value of *req* transitioned to TRUE while waiting for an acknowledge or while acknowledge was asserted. Extra requests do not initiate new transactions.
ACK_WITHOUT_REQ_ VIOLATION	The value of *ack* transitioned to TRUE without a pending request.
ACK_MIN_CYCLE_ VIOLATION	The value of ack transitioned to TRUE before *min_ack_cycle* clock cycles transpired after the request.
ACK_MAX_CYCLE_ VIOLATION	The value of *ack* did not transition to TRUE before *max_ack_cycle* clock cycles transpired after the request.
REQ_DROP_VIOLATION	The value of *req* transitioned from TRUE before an acknowledge.
REQ_DEASSERT_VIOLATION	The value of req did not transition from TRUE before *deassert_count* clock cycles transpired after an acknowledge.
ACK_MAX_LENGTH_ VIOLATION	The value of ack did not transition from TRUE before *max_ack_length* clock cycles transpired after an acknowledge.

FIGURE 7-10. Assertions in `ovl_handshake`

Here is `ovl_handshake` failing the Questa waveform tool:

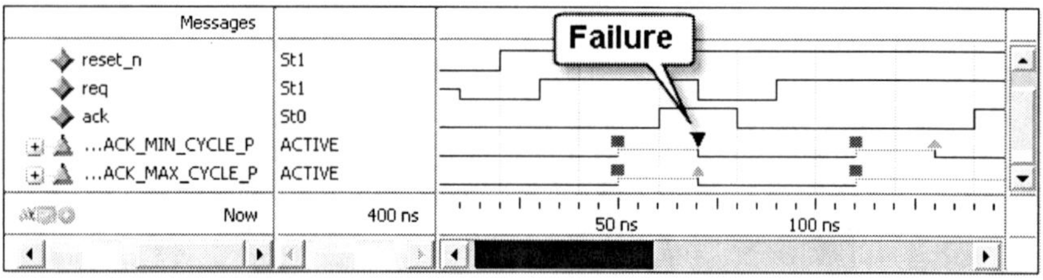

FIGURE 7-11. Running `ovl_handshake`

There are two assertions here. The failure happens on the assertion that checked the minimum number of cycles. The assertion checking the maximum number of cycles had no problem with an early response and passed.

We've instantiated an immediate assertion and a multicycle assertion in a Verilog assertion module. Now we need to see how to compile these modules.

7.3 Compiling with the OVL

In our previous examples we used two OVL checker modules: `ovl_never` and `ovl_handshake`. These are module names in the Open Verification Library. In order to use them, we need to tell the Verilog compiler where to find them.[6]

The OVL is stored in a directory called `std_ovl`. This directory contains the module definitions for all the checkers. It also contains an include file that defines all the OVL macros.

When we compile the OVL we need to tell the Verilog compiler four things:

- Where to find the OVL library.
- The extension to add to a module name to covert it to a filename.
- Where to find the OVL include files.
- Which macros are defined while compiling.

We store all of this information in a `.f` file that we'll reference with our Verilog compiler command. We'll look at the compiler options that go into the `.f` file and we'll view that file at the end of the discussion.

7.3.1 The OVL Library Information

The `-y` lets us specify a directory that contains library modules. In the case of the OVL, we point to `std_ovl`. In our examples, we assume that `std_ovl` is two directory levels above our code (which it is in all the examples on www.fpgasimulation.com.)

We need to tell the compiler where to find the source files and what extension to add to the module names to create the file names. The options look like this:

```
-y ../../std_ovl
   +libext+.v
```

6. This chapter assumes that you are familiar with the concepts in Chapter 6.

7.3.2 The OVL Include File Information

The file that defines our assertion module needs to include a header file called `std_ovl_defines.h`. This file defines all the macros for OVL compilation. This file is also stored in `std_ovl`, so we use the `+incdir` option to share this fact with the Verilog compiler when we compile the assertion module:

```
+incdir+../../std_ovl
```

The above line works with the example files that can be downloaded from www.fpgasimulation.com. This line says that the `std_ovl` directory is two directory levels above the current directory.

7.3.3 OVL Control Macros

We control the OVL compilation with macros. We need to define two macros whenever we compile the `assertion` module:

- `OVL_ASSERT_ON`—We define this macro whenever we want to compile or synthesize with assertions.

  If we don't want macros for a given simulation or synthesis run, we simply leave `OVL_ASSERT_ON` undefined and the Verilog compiler never sees the assertion code.
- `OVL_SVA` or `OVL_VERILOG`—These macros tell the OVL which language to use. We define only one of them. If we define `OVL_SVA,` we'll have access to all 50 of the checkers (in OVL 2.3.) If we define `OVL_VERILOG,` we'll have access to 33. Table 2-1 in the OVL LRM tells us which checkers exist for which languages.

We define the macros to use SystemVerilog Assertions as follows:

```
+define+OVL_ASSERT_ON +define+OVL_SVA
```

If we want to use the Verilog OVL, we define the macros as follows:

```
+define+OVL_ASSERT_ON +define+OVL_VERILOG
```

7.3.4 The Compiler Option File

The Verilog compiler allows us to capture all our compiler options in a file and pass them to the compiler with the `-f` option. By convention, I give these files a `.f` extension.

We compile a file using the `-f` option like this:[7]

```
vlog -f compile.f
```

The compilation command doesn't need any more arguments, because these are stored in the `compile.f` file:

```
+define+OVL_SVA
+define+OVL_ASSERT_ON

-y ../../std_ovl
+libext+.v

+incdir+../../std_ovl
```

FIGURE 7-12. The compile.f File for the `assertion` Module

This file implements the command line options and finds the OVL checker modules.

7.4 Summary

In this chapter, we instantiated OVL checkers in a Verilog module.

The strategy of encapsulating all the modules in a Verilog module makes our lives much easier in a mixed-language environment. It allows us to drop the Verilog checkers into a VHDL design without having to worry about the issues associated with mixing the Verilog and VHDL library management processes.

We saw that the LRM shows us everything we need to know in order to instantiate a checker, and we saw how to use parameters to control the checker.

Finally, we looked at the options we need to invoke the OVL:

```
-y, +libext, +incdir, +define
```

7. `vlog` is the Questa command. Incisive has `ncvlog` and Riviera has `alog`.

We put these options into a compile file and invoked them on the Verilog compiler with the `-f` option.

The SystemVerilog Assertion version of the OVL checkers is the largest and most robust library. However, not everyone can access these checkers. If we don't have access to a mixed-language simulator, or if we have a project requirement that allows only VHDL code, we need to use the VHDL version of the Open Verification Library.

Because VHDL lacks include files or macros, and because it has a completely different library-management mechanism, we cannot use what we've learned about Verilog in a VHDL implementation. Using the OVL with VHDL is the subject of our next chapter.

8 Using the VHDL OVL

IN THIS CHAPTER:

- *VHDL Library Management*
- *Compiling the VHDL OVL*
- *Instantiating VHDL Checkers*
- *The OVL Control Record*

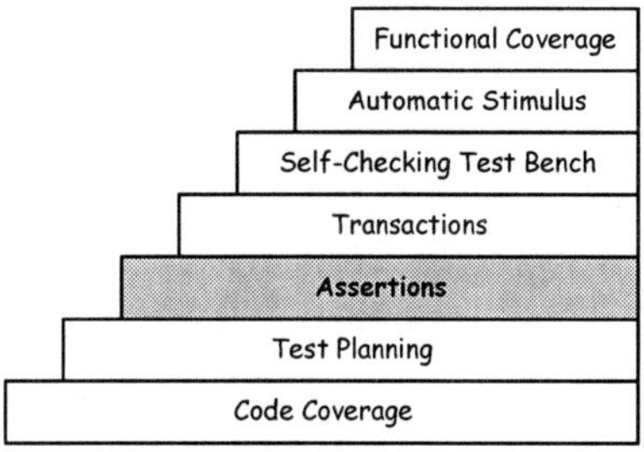

It is a basic tenet of biology that no two creatures fill the same niche in an ecosystem. One competitor always snuffs out another. Squirrels and chipmunks are both rodents that live in the forest, and both can climb trees, but squirrels live high in the trees while chipmunks live underground with moles. The moles live underground, but they eat different things than chipmunks and nature's balance is maintained.

Not so with VHDL and Verilog. These languages coexist in the same niche because they were created at about the same time, and both wound up with a body of intellectual property that cannot be thrown away. As a result, we are forced to learn how to do everything twice: once in Verilog and again in VHDL.[1]

Today's simulators can help us manage this situation by simulating both languages in a single simulation run. This means we can use the full range of OVL assertions in a VHDL design by creating a single Verilog assertion module and instantiating Verilog or SystemVerilog checkers.

However, simulator or project restrictions may force us to use nothing but VHDL. If this happens we should not give up the idea of using OVL checkers.

1. Or the other way around.

They are far too valuable to be tossed aside. Instead, we should use the VHDL version of the OVL. This library delivers ten checkers written entirely in VHDL:

TABLE 8-1. OVL VHDL Checkers

`ovl_always`	`ovl_cycle_sequence`
`ovl_implication`	`ovl_never`
`ovl_never_unknown`	`ovl_never_unknown_async`
`ovl_next`	`ovl_one_hot`
`ovl_range`	`ovl_zero_one_hot`

These are the most popular checkers in the OVL, so we can get a lot done despite having a limited number of checkers available to us.

8.1 VHDL Library Management

VHDL and Verilog have fundamentally different ways of approaching library management. Verilog follows the C model where a collection of include files, macros, and conditional compilations control what gets compiled. This means that you recompile your design and library when making a change.

VHDL does things differently. It requires that we compile everything into a library, and then we control what gets used. The advantage of the VHDL approach is that we can compile a central OVL library that all users can access.

We need to know how to do three things to use the VHDL OVL:

1. Compile the OVL into two libraries called `accellera_ovl_vhdl` and `accellera_ovl_library_vhdl_u`. (The "u" version has unsigned standard logic numbers.)
2. Instantiate the checkers.
3. Create a control record if we want behavior other than the default.

We'll examine each of these steps, using our friend the three-bit counter as our example.

8.2 Compiling the VHDL OVL

The VHDL OVL assumes that we have compiled two libraries to store the OVL components: `accellera_ovl_vhdl` and `accellera_ovl_vhdl_u`. Simulation companies that supply precompiled OVL VHDL libraries use these library names.

We compile the files into `accellera_ovl_vhdl` in the following order:

1. `std_ovl/std_ovl.vhd`
2. `std_ovl/std_ovl_components.vhd`
3. `std_ovl/std_ovl_procs.vhd`
4. `std_ovl/std_ovl_clock_gating.vhd`
5. `std_ovl/std_ovl_reset_gating.vhd`
6. `std_ovl/ovl_*.vhd`
7. `std_ovl/vhdl93/ovl_*_rtl.vhd`

Once we've done this, our `std_logic` OVL checkers are ready. If we want `std_ulogic` checkers, we need to compile the `std_ovl_u_components.vhd` (notice the "`u`" in that filename) into `accellera_ovl_vhdl_u`.

Here is a Tcl script that compiles the VHDL OVL using Questa:

```
1 if {[file exists accellera_ovl_vhdl]==0} {
2   vlib accellera_ovl_vhdl
3   vcom -work accellera_ovl_vhdl -f compile_ovl.f
4   vlib accellera_ovl_vhdl_u
5   vcom -work accellera_ovl_vhdl_u ../../std_ovl/std_ovl_u_components.vhd
6 }
```

FIGURE 8-1. VHDL OVL Compilation Script

One advantage of VHDL is that we can compile the OVL library once and keep reusing it. Consequently, this Tcl script checks for the existence of `accellera_ovl_vhdl` and compiles only if that library is missing.

Line 3 does the actual compilation. This line takes advantage of the `-f` option in the `vcom` command to store all the filenames in a file called `compile_ovl.f`, rather than typing them all on the command line or running `vcom` many times.

Here is `compile_ovl.f`. This file works for all OVL compilations:[2]

```
../../std_ovl/std_ovl.vhd
../../std_ovl/std_ovl_components.vhd
../../std_ovl/std_ovl_procs.vhd
../../std_ovl/std_ovl_clock_gating.vhd
../../std_ovl/std_ovl_reset_gating.vhd

../../std_ovl/ovl_always.vhd
../../std_ovl/ovl_cycle_sequence.vhd
../../std_ovl/ovl_implication.vhd
../../std_ovl/ovl_never.vhd
../../std_ovl/ovl_never_unknown.vhd
../../std_ovl/ovl_never_unknown_async.vhd
../../std_ovl/ovl_next.vhd
../../std_ovl/ovl_one_hot.vhd
../../std_ovl/ovl_range.vhd
../../std_ovl/ovl_zero_one_hot.vhd

../../std_ovl/vhdl93/ovl_always_rtl.vhd
../../std_ovl/vhdl93/ovl_cycle_sequence_rtl.vhd
../../std_ovl/vhdl93/ovl_implication_rtl.vhd
../../std_ovl/vhdl93/ovl_never_rtl.vhd
../../std_ovl/vhdl93/ovl_never_unknown_async_rtl.vhd
../../std_ovl/vhdl93/ovl_never_unknown_rtl.vhd
../../std_ovl/vhdl93/ovl_next_rtl.vhd
../../std_ovl/vhdl93/ovl_one_hot_rtl.vhd
../../std_ovl/vhdl93/ovl_range_rtl.vhd
../../std_ovl/vhdl93/ovl_zero_one_hot_rtl.vhd
```

FIGURE 8-2. VHDL OVL Source File List

Now that we have the `accellera_ovl_vhdl` library in place, we can instantiate checkers.

8.3 Instantiating VHDL Checkers

As we did with Verilog-based checkers, we are going to store all our checkers in a single component and instantiate that component either as a firewall or as a protocol checker.

In this example, we are going to create the firewall for the three-bit counter. We don't want this counter to overflow, so if its output is `3'h7` and the increment signal is high, then we signal an error because this would create an overflow.

2. You can download these example files from www.fpgasimulation.com.

Here is the firewall component for the three-bit counter, written in VHDL. This module is identical to the firewall module we implemented in Verilog in Figure 7-4 on page 80. It takes `data_out` and the `inc` signal as input and fires an assertion if we try to increment when the counter is full:

```
1  library ieee;
2  use ieee.std_logic_1164.all;
3  use ieee.std_logic_arith.all;
4  library accellera_ovl_vhdl;
5  use accellera_ovl_vhdl.std_ovl.all;
6  use accellera_ovl_vhdl.std_ovl_components.all;
7
8  entity threebitcounter_assert is
9
10   port (
11     clk          : in  std_logic;
12     rst          : in  std_logic;
13     inc          : in  std_logic;
14     ld           : in  std_logic;
15     data_in      : in  std_logic_vector(2 downto 0);
16     data_out     : in  std_logic_vector(2 downto 0)
17     );
18
19 end threebitcounter_assert;
20
21 architecture RTL of threebitcounter_assert is
22
23   signal overflow : std_logic;
24   signal fire :     std_logic_vector(OVL_FIRE_WIDTH-1 downto 0);
25   signal enable:    std_logic;
26
27 begin  -- RTL
28
29   overflow <= '1' when ((data_out = "111") and (inc = '1')) else '0';
30   enable   <= '1' when (rst = '0') else '1';
31
32   U_counter_overflow: ovl_never
33     generic map (
34       reset_polarity => OVL_ACTIVE_HIGH,
35       msg            => "COUNTER OVERFLOW!!!! OH NO!!!")
36     port map (
37       clock     => clk,
38       reset     => rst,
39       enable    => enable,
40       test_expr => overflow,
41       fire      => fire);
42
43 end RTL;
```

FIGURE 8-3. VHDL Firewall Assertion Module for `threebitcounter`

- **Lines 1–6**—VHDL uses libraries to access common elements rather than `include` statements.
- **Lines 8–18**—This is the entity declaration for our assertion component. The component only has inputs, because we are simply monitoring data.
- **Line 21**—We have a single architecture for this entity.
- **Line 23**—VHDL does not allow us to put an expression on a port, so we need to create a signal to capture the expression that should never be true.

- **Line 24**—The checker has an output bus that we need to support.
- **Line 25**—The `enable` signal is the inversion of the `rst` signal.
- **Line 27**—The `overflow` signal becomes true if `data_out` is at `3'h7` and the `inc` signal is high.
- **Line 30**—Invert the `rst` signal and store it in `enable`.
- **Lines 32–41**—We instantiate the `ovl_never` checker and call it `U_counter_overflow`.
- **Line 34**—The design has an active high reset, so we set the checker to use an active high reset.
- **Line 35**—We replace the default message with one that makes us smile.
- **Lines 37–41**—Hook up the signals.

In the end, using VHDL checkers boils down to the same thing as using Verilog checkers. You pick your checker and instantiate it in an assertion component.

8.4 The OVL Control Record

In Verilog, we controlled the OVL with macros that conditionally enabled and disabled chunks of code. We can't do that in VHDL. Instead, we compile the entire library into the `accellera_ovl_vhdl` library and then control the library with generics.

All VHDL OVL checkers have a generic that is unique to VHDL. This is the `controls` generic. We use the generic by passing it an `ovl_ctrl_record`.

The OVL defines the `ovl_ctrl_record` type in the `std_ovl.vhd` file. This record contains OVL control information. The OVL also defines a default control-record constant called `ovl_ctrl_defaults`. The OVL uses this default record whenever a user instantiates a checker without specifying a value for the `controls` generic.

We saw an example of a default instantiation in Figure 8-3 on page 95, where we instantiated `ovl_never` and ignored the controls generic.

If we want to modify the defaults, we need to create our own `ovl_ctrl_record` constant and pass it to the checkers. The easiest way to do that is to open the `std_ovl.vhd` file, copy the default record from there, and modify it.

Here's the default control record `ovl_ctrl_defaults`:

```
constant OVL_CTRL_DEFAULTS                        : ovl_ctrl_record
(
  -- generate statement controls
  xcheck_ctrl                  => OVL_ON,
  implicit_xcheck_ctrl         => OVL_ON,
  init_msg_ctrl                => OVL_OFF,
  init_count_ctrl              => OVL_OFF,
  assert_ctrl                  => OVL_ON,
  cover_ctrl                   => OVL_OFF,
  global_reset_ctrl            => OVL_OFF,
  finish_ctrl                  => OVL_ON,
  gating_ctrl                  => OVL_ON,

  -- user configurable library constants
  max_report_error             => 15,
  max_report_cover_point       => 15,
  runtime_after_fatal          => "100 ns     ",

  -- default values for common generics
  severity_level_default       => OVL_SEVERITY_DEFAULT,
  property_type_default        => OVL_PROPERTY_DEFAULT,
  msg_default                  => OVL_MSG_DEFAULT,
  coverage_level_default       => OVL_COVER_DEFAULT,
  clock_edge_default           => OVL_CLOCK_EDGE_DEFAULT,
  reset_polarity_default       => OVL_RESET_POLARITY_DEFAULT,
  gating_type_default          => OVL_GATING_TYPE_DEFAULT
);
```

FIGURE 8-4. Definition of `ovl_ctrl_defaults` in `std_ovl.vhd`

The file `ovl_lrm.pdf` describes these controls. The one that we would use the most is `assert_ctrl`, which turns assertions off and on.

Let's say we want to change three things in our checker behavior:

- The number of errors we'll report
- The amount of time between a fatal error and stopping the simulator
- The default error message

We do this by copying the default control record out of `std_ovl.vhd`, and creating our own control record file. Then we create our own constant value of type `ovl_ctrl_record`. We'll call our control record `my_proj_controls`.

We create a file called `control_record.vhd` and define the `my_ovl_control_pkg` package. We paste a copy of the default control record into the package and modify it like this:

```
1  library accellera_ovl_vhdl;
2  use accellera_ovl_vhdl.std_ovl.all;
3  use accellera_ovl_vhdl.std_ovl_procs.all;
4
5  package my_ovl_control_pkg is
6  -- OVL configuration
7    constant my_ovl_controls : ovl_ctrl_record := ( -- renamed constant
8  -- generate statement controls                  -- from ovl_ctrl_default
9       xcheck_ctrl            =>     OVL_ON,
10      implicit_xcheck_ctrl   =>     OVL_ON,
11      init_msg_ctrl          =>     OVL_OFF,
12      init_count_ctrl        =>     OVL_OFF,
13      assert_ctrl            =>     OVL_ON,
14      cover_ctrl             =>     OVL_OFF,
15      global_reset_ctrl      =>     OVL_OFF,
16      finish_ctrl            =>     OVL_ON,
17      gating_ctrl            =>     OVL_ON,
18
19 -- user configurable library constants
20      max_report_error       =>     4,   -- changed
21      max_report_cover_point =>     15,
22      runtime_after_fatal    =>    "150 ns     ", -- changed
23
24 -- default values for common generics
25      severity_level_default => OVL_SEVERITY_DEFAULT,
26      property_type_default  => OVL_PROPERTY_DEFAULT,
27      msg_default            => ovl_set_msg("XXXXX ERROR"), -- changed
28      coverage_level_default => OVL_COVER_DEFAULT,
29      clock_edge_default     => OVL_CLOCK_EDGE_DEFAULT,
30      reset_polarity_default => OVL_ACTIVE_HIGH,
31      gating_type_default    => OVL_GATING_TYPE_DEFAULT
32      );
33 end package my_ovl_control_pkg;
```

FIGURE 8-5. Customized Control Record

We've modified four spots in this record:

- **Line 7**—We renamed the constant to `my_ovl_controls`.
- **Line 20**—We changed the maximum report error from 4.
- **Line 22**—We tell the simulation to run for 150 ns after it detects a fatal error.
- **Line 27**—We changed the message so it has a string that makes it easy to find errors with a script.

To use the record, we compile `control_record.vhd` with the rest of our source code, so that we have the package available. Then we reference the package and pass the control record to our OVL checkers with the `controls` generic.

Here's our friend, the three-bit counter assertion, using our control record:

```
1  library ieee;
2  use ieee.std_logic_1164.all;
3  use ieee.std_logic_arith.all;
4  library accellera_ovl_vhdl;
5  use accellera_ovl_vhdl.std_ovl.all;
6  use accellera_ovl_vhdl.std_ovl_components.all;
7
8  use work.my_ovl_control_pkg.all;
9
10 entity threebitcounter_assert is
11
12   port (
13     clk          : in  std_logic;
14     rst          : in  std_logic;
...
35   u_counter_overflow: ovl_never
36     generic map (
37       reset_polarity => OVL_ACTIVE_HIGH,
38       controls       => my_ovl_controls)
39     port map (
40       clock     => clk,
41       reset     => rst,
42       enable    => enable,
43       test_expr => overflow,
44       fire      => fire);
45
46 end RTL;
```

FIGURE 8-6. Using a Control Record

We need to reference our project package to see the control record. Then we need to use the generic to pass the control record:

- **Line 8**—This `use` command references the `my_ovl_control_pkg` package.
- **Line 37**—This generic line passes our control record constant, `my_ovl_controls`, to the checker.

Now when we simulate we'll be able to control all our OVL checkers by changing the control record and recompiling it.

8.5 Summary

In this chapter, we learned how to use the OVL when we have access to only a VHDL simulator or if the project forces us to use only VHDL. If we have access to a mixed-language simulation, then we should use the Verilog or SystemVerilog Assertion checkers so as to have more flexibility.

We saw that the VHDL OVL assumes that we compiled all the VHDL checkers into two libraries: `accellera_ovl_vhdl` and `accellera_ovl_vhdl_u`. These libraries need to be compiled only once, and they can be centrally located for a project team.

We control the VHDL OVL checkers with an `ovl_control_record`. The OVL defines this type and provides a default. If we want to override the default, we need to compile our own `ovl_control_record` constant and pass it to the OVL checkers with the `controls` generic.

Now we can simulate VHDL or Verilog designs with assertions that catch bugs as soon as they happen. The assertions write a message that lets us stop our simulation when things go badly.

Next, we're going catch bugs as they happen in the lab. Many of the OVL assertions are synthesizable, creating signals in our chip that can trigger a logic analyzer to pinpoint bugs at board speeds. We'll learn how to synthesize checkers in the next chapter.

9 Using Assertions in the Lab

IN THIS CHAPTER:

- *Synthesizable OVL Checkers*
- *The* `error` *Signal*
- *Using Multiple Checkers in One Block*
- *Building the Error Tree*
- *Removing Assertion Hardware*

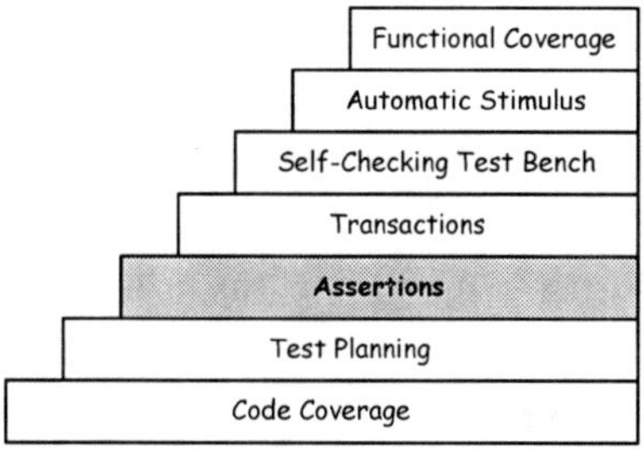

Debugging is an experience in time travel. When we debug, we look at the bad output in our design and travel back in time to find out what went wrong and how. Unlike science fiction time travel, debugging time travel becomes more difficult the further back you go. A bug whose source happened a hundred clocks ago is more difficult to find than one that happened two clocks ago.

The problem of time travel becomes even more difficult in a lab debug session. The design speeds and lack of visibility make it very hard to capture the signals that we need to debug a problem.

What we'd really like to have is a board-mounted LED that goes red when there's a bug in our design. Then we could hook the logic analyzer to it and read the signals just before the bug occurred. Better yet, we'd like to be able to trace the signal to that LED back into the chip and find the exact block that caused the problem. This would make lab debug much easier.

We have the power to do this today. By using the synthesizable OVL checkers, we can create a signal that will tell us when our design has gone into the weeds, and we can trace that signal down through the chip to find the exact spot where something has gone wrong.

9.1 Synthesizable OVL Checkers

There are eight synthesizable checkers. They are available in both Verilog and VHDL except for `ovl_win_unchange`, which is available only in Verilog.

TABLE 9-1. Synthesizable Checkers in OVL V2.3

ovl_always	ovl_cycle_sequence
ovl_implication	ovl_never
ovl_never_unknown	ovl_next
ovl_one_hot	ovl_range
ovl_win_unchange	ovl_zero_one_hot

Checkers other than these can be run through a synthesizer, but their error signals are tied to zero.

9.1.1 The OVL Error Signal

All OVL checkers have an output bus called `fire`. In our previous examples we connected to this bus but never used it. We'll get to use it now.

The bit `fire[0]` is the error signal for synthesizable checkers. Because this signal cannot be `X` or `Z`, it is called the two-state output signal. We access it like this:

```
fire['OVL_FIRE_2STATE] // Verilog

fire(OVL_FIRE_2STATE) -- VHDL
```

We will use this signal when we create our new assertion module with an `error` output. When we synthesize checkers we use the Verilog or VHDL versions, as not all synthesis tools can read SystemVerilog.

OVL checkers have unsynthesizable code that writes messages to the screen. The checkers handle this code in two ways, depending upon whether you are using VHDL or Verilog.

9.1.2 Synthesizing Verilog Checkers

The Verilog OVL disables unsynthesizable code by using conditional compilation. Defining the `OVL_SYNTHESIS` macro hides all the unsynthesizable code from the synthesis tool.

Here's how we define the macros in the Precision RTL[1] synthesis tool.

```
setup_design -defines {+define+OVL_ASSERT_ON
 +define+OVL_VERILOG +define+OVL_SYNTHESIS}
```

We've used the `OVL_ASSERT_ON` and `OVL_VERILOG` macros before. The new macro is `OVL_SYNTHESIS`. When this is defined we don't see any messages, but we can get an error signal.

9.1.3 Synthesizing VHDL Checkers

VHDL cannot hide code the way Verilog can, so when we compile VHDL for synthesis we need to use different files to define the architectures. We are still compiling into the `accellera_ovl_vhdl` library.

We compile the files in the following order. The differences between this compilation and the one for simulation are in boldface:

1. `std_ovl/std_ovl.vhd`
2. `std_ovl/std_ovl_components.vhd`
3. `std_ovl/`**`vhdl93/syn_src`**`/std_ovl_procs.vhd`
4. `std_ovl/std_ovl_clock_gating.vhd`
5. `std_ovl/std_ovl_reset_gating.vhd`
6. `std_ovl/ovl_*.vhd`
7. `std_ovl/vhdl93/`**`syn_src`**`/ovl_*_rtl.vhd`

Even with these changes, the VHDL in the `syn_src` directory may not work in all synthesis tools. In those cases, you'll need to make your own copy of the VHDL architectures and remove references to procedures such as `ovl_init_msg_proc`.

1. Precision RTL is an FPGA synthesis tool available from Mentor Graphics.

9.2 The `error` Signal

Your OVL checkers should all be stored in a block called an assertion block. As we've seen before, there are two kinds of assertion blocks: firewall blocks and protocol monitor blocks.

There are many advantages to storing all the checkers in a single block. The block makes it easier to manage mixed-language simulation. People reusing our design can remove the assertions easily, and we know where to go if we want to add our assertion signals to our waveform.

Now we have a new advantage for the assertion block. It can generate a signal called `error` that we can bring out to the top level of our chip and attach to a logic analyzer or LED. We create the error signal by adding an output to our assertion block.

For example, here is the architecture for our three-bit counter's new assertion block. The block is identical to our previous block, but now we're actually using the `fire` bus that comes out of the OVL checker.

The fire bus has three signals:

- **`fire(0)`**—Set if there is an error.
- **`fire(1)`**—Set if there is an X or Z in the checker.
- **`fire(2)`**—Set if we've triggered a coverage point.[2]

We're interested only in the `fire(0)` signal, but rather than hard code the number 0 into the index, we'll use the OVL constant `OVL_FIRE_2STATE` as an index into the fire `bus`. We'll attach the `fire(OVL_FIRE_2STATE)` signal to the error signal and bring the error signal out of the assertion block. Then we'll bring that signal all the way up to the top of our design.

2. Discussed in Chapter 27

Here is the source code:

```
24 architecture RTL of threebitcounter_assert is
25
26   signal overflow : std_logic;          -- capture the overflow condition
27   signal fire : std_logic_vector(OVL_FIRE_WIDTH-1 downto 0);  -- assertion
28
29   COMPONENT ovl_never
30     GENERIC (
31       msg            : string              := OVL_MSG_NOT_SET;
32       controls       : ovl_ctrl_record     := OVL_CTRL_DEFAULTS;
33       reset_polarity : ovl_reset_polarity := OVL_RESET_POLARITY_NOT_SET
34       );
35     PORT (
36       clock     : IN     std_logic;
37       reset     : IN     std_logic;
38       enable    : IN     std_logic;
39       test_expr : IN     std_logic;
40       fire      : OUT    std_logic_vector(OVL_FIRE_WIDTH - 1 downto 0)
41       );
42    end COMPONENT;
43
44 begin  -- RTL
45
46   overflow <= '1' when ((data_out = "111") and (inc = '1')) else '0';
47
48   error <= fire(OVL_FIRE_2STATE);
49
50   cntr_overflow: ovl_never
51     generic map (
52       reset_polarity => OVL_ACTIVE_HIGH,
53       controls       => ovl_proj_controls)
54     port map (
55       clock     => clk,
56       reset     => rst,
57       enable    => '1',
58       test_expr => overflow,
59       fire      => fire);
60
61 end RTL;
```

FIGURE 9-1. VHDL Assertion Block with error Signal

- **Line 24**—Create a component to hold our checkers. In this case our block contains only the ovl_never checker.
- **Line 27**—Create a bus to hold the fire outputs from ovl_never.
- **Line 40**—The ovl_never checker outputs a bus of three signals.
- **Line 48**—We attached the error signal to the fire(OVL_FIRE_2STATE) signal from the ovl_never checker.
- **Line 59**— Attach the fire bus at this level to the fire output from ovl_never.

Here is the synthesized result of this block in the Precision RTL schematic viewer:

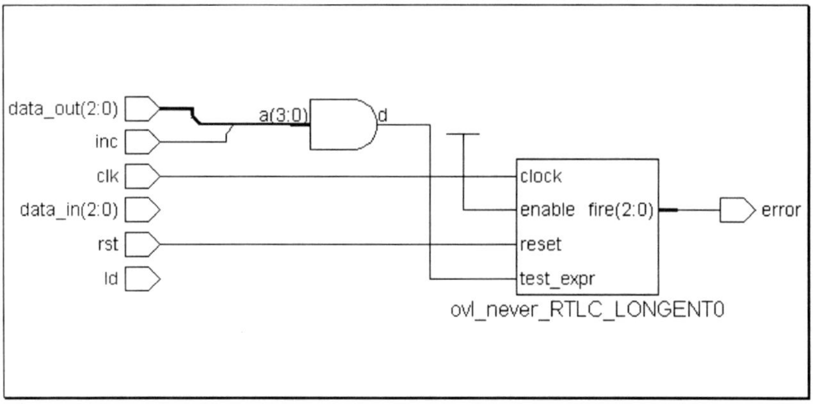

FIGURE 9-2. Synthesized `ovl_never` Assertion

We can now see our `ovl_never` synthesized into real hardware. We said that the `inc` signal should never be high when the output is `3'b111`. This translates into a four-bit AND gate. The AND gate feeds the output flop, which drives the `error` signal.

You can also see the synthesis tool working to reduce the hardware in this design. Although we passed all the signals from the counter into the assertion module, we never used the `data_in(2:0)` or `ld` signals. These are optimized away.

This is all hidden from the top level of the three-bit counter. As far as the top level is concerned, it has an assertion block that takes all the signals and produces a single output. As we'll see later, that single output stays the same no matter how many checkers we add to the block. This is another advantage of encapsulating all the checkers in a single module.

As we see below, the top level of the three-bit counter includes a counter whose `updn` pin is tied high. There are two output ports, the counter output and the error signal.

Here is the schematic of the counter with the assertion module:

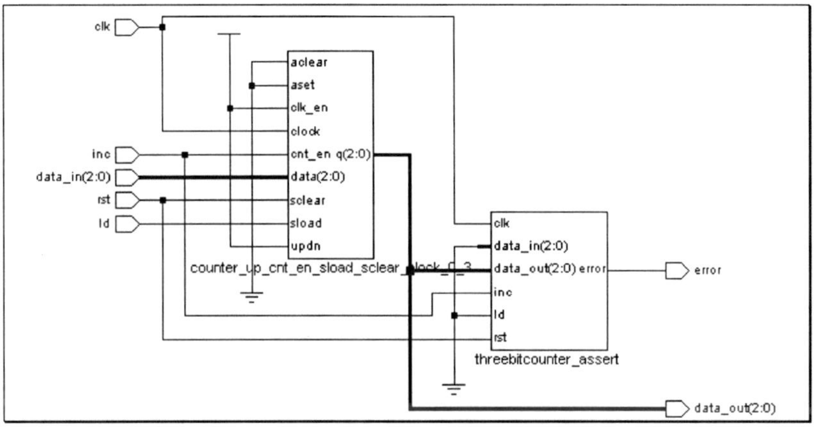

FIGURE 9-3. Top-Level Assertion Output

In this example we saw one checker in one module creating one error signal. In our next section we'll add multiple checkers, but still make one output signal.

9.3 Multiple Checkers in One Block

The three-bit counter is a tiny example, and so it can be checked completely with one checker. That is unusual. The typical case will have multiple checkers in one block. That's where creating an `assertion` module adds even more value.

We implement multiple checkers by creating a `fire` bus for each checker in the design and then ORing the `fire['OVL_FIRE_2STATE]` signals from all the checkers into one `error` output.

As an example, let's add another checker to the three-bit counter. We'll add the requirement that we cannot load data into the counter when we are incrementing it. This means that the `ld` signal needs to be low if the `inc` signal is high.

We'll check this requirement with the `ovl_implication` checker. This checker says that if the antecedent expression is true, then the consequent expression must also be true. In this case the antecedent expression is the signal `inc`, and the consequent expression is the inversion of `ld` (`~ld`).

Here is our new assertion module. We are going to create a Verilog assertion module, since this is the most flexible way to use the OVL. This module shows how we can add multiple checkers to a design and create one error signal:

```
1  `include "std_ovl_defines.h"
2  module assertion (input clk, input rst, input ld, input inc,
3                    input [2:0] data_in, input [2:0] data_out,
4                    output error);
5
6  wire [`OVL_FIRE_WIDTH-1:0] overflow_fire,
7                             ld_on_inc_fire;
8
9  assign error = overflow_fire[`OVL_FIRE_2STATE] |
10                ld_on_inc_fire[`OVL_FIRE_2STATE];
11
12
13 ovl_never #(.reset_polarity(`OVL_ACTIVE_HIGH))
14
15 U_counter_overflow
16          (.clock(clk), .reset(rst), .enable(~rst), .fire(overflow_fire),
17
18           .test_expr((data_out == 3'h7) && inc));
19
20
21
22 ovl_implication#(.reset_polarity(`OVL_ACTIVE_HIGH))
23
24 U_no_ld_on_inc
25          (.clock(clk), .reset(rst), .enable(~rst), .fire(ld_on_inc_fire),
26
27          .antecedent_expr(inc),
28          .consequent_expr(~ld));
29
30 endmodule
```

FIGURE 9-4. Multi-Checker `assertion` Module

- **Line 4**—Declare a single-bit output called `error`.
- **Lines 6–7**—Declare an independent fire bus for each checker in the module. We have two checkers, so we have two fire buses.
- **Lines 9–10**—OR the `` `OVL_FIRE_2STATE `` bit from all the fire buses together to create a single error signal.
- **Lines 13–18**—Instantiate the `ovl_never` checker.
- **Lines 22–28**—Instantiate the `ovl_implication` checker.
- **Line 27**—Our antecedent expression is the `inc` signal.
- **Line 28**—Our consequent expression is the inversion of `ld`.

This gives us an assertion module with one error signal. Here's what it looks like in the synthesis tool:

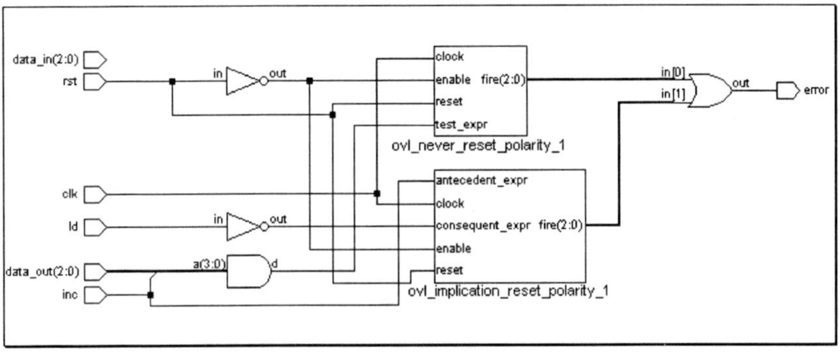

FIGURE 9-5. Assertion Module with Two Checkers

Now that we've seen how to add multiple checkers to a module, let's add multiple modules to a design.

9.4 Building an Error Tree

Once we have output signals on all our blocks, our next job is to get those signals to the top of the design. Ultimately we want an error signal that we can tie to an LED to tell us when we've had an assertion fire.

In the preceding section, we pulled multiple checkers together to create an error signal from the assertion module. Now we'll pull multiple error signals together to make one error signal at the top of our block.

There are many ways to bring error signals to the top of a chip. We will focus on the simplest of them here by creating an OR tree. There are more-complex approaches that, for example, encode all the assertion signals so that each one has a unique number.[3]

3. Go to www.fpgasimulation.com for articles on these techniques.

Our approach is to create a big OR tree that results in single error signal at the top level. If there is a bug, we use the inputs to the OR gate to find out which block had the error, and then work our way down the tree using FPGA debug tools to bring error signals to the output pins.

9.4.1 Combining Block Error Signals

Our first step is to combine the error signals from our blocks into a single output signal. We do this by creating an OR reduction of all the error signals. In this example, we created an error bus and then did an OR reduction on the bus.

Each instance of `threebitcounter` has an error output, so we've attached each error output to a different bit in the error bus.:

```
1  module threecounters(
2      input clk,
3      input rst,
4      input inc,
5      input ld,
6      input [2:0] data_in,
7      output[2:0] data_out,
8      output error);
9
10     wire [2:0] data_out1, data_out2;
11     wire [1:3] error_bus;
12
13     assign error = |error_bus;
14
15     threebitcounter U1 (
16            .clk(clk), .rst(rst), .inc(inc), .ld(ld), .data_in(data_in),
17            .data_out(data_out1), .error(error_bus[1]));
18
19     threebitcounter U2 (
20            .clk(clk), .rst(rst), .inc(inc), .ld(ld), .data_in(data_out1),
21            .data_out(data_out2), .error(error_bus[2]));
22
23     threebitcounter U3 (
24            .clk(clk), .rst(rst), .inc(inc), .ld(ld), .data_in(data_out2),
25            .data_out(data_out ), .error(error_bus[3]));
26
27   endmodule
```

FIGURE 9-6. Combining Error Signals Among Blocks

- **Line 8**—Create a single `error` signal that will be set if any of the checkers fire.
- **Line 11**—Define a bus with one signal per checker.
- **Line 13**—Perform an OR reduction on the error signals from the three checkers.
- **Line 17**—Attach `U1` to bit-1 on `error_bus`.
- **Line 21**—Attach checker `U2` to bit-2 on `error_bus`.
- **Line 25**—Attach checker `U3` to bit-3 on `error_bus`.

This code looks like this after being synthesized in Precision RTL:

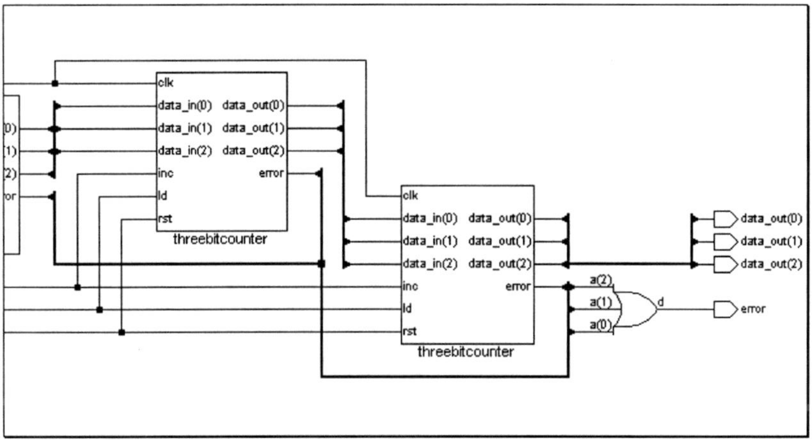

FIGURE 9-7. Three Error Signals Reduced by an OR Gate

We can make this OR Gate as wide as we like, because we aren't worried about the performance of our error tree. We'll see later that it is easy to remove the extra logic when we are done debugging.

9.4.2 Creating the Top Level

We can build our OR tree to be as many levels as we like, with each level combining the errors among its blocks. Eventually we will need to reach the top-level. When we do reach the top level, we want to capture our error signal in a flip-flop that will hold the signal until we reset the design. This way we can catch a fleeting error.

We can create the set of error signals by creating an error bus as we did in Section 9.4.1, or we can give each block its own signal. In cases where we have just a few blocks at the top level, we may want to give the error signals meaningful names associated with the block.

In our example, we'll take two of the `threebitcounter` modules and combine them into a top level.

The source code looks like this:

```
1  module sixcounters(
2     input clk,rst, ld, inc, select,
3     input [2:0] data_in,
4     output[2:0] data_out,
5     output reg error);
6
7     wire [2:0] data_out0;
8     wire       error0, error1;
9
10    always @(posedge clk)
11       if (!rst)
12          error <= 0;
13       else
14          if (!error) error <= error0 | error1;
15
16
17    threecounters U0 (
18       .clk(clk), .rst(rst), .ld(ld), .inc(inc), .data_in(data_in),
19       .data_out(data_out0), .error(error0));
20
21    threecounters U1(
22       .clk(clk), .rst(rst), .ld(ld), .inc(inc), .data_in(data_out0),
23       .data_out(data_out ), .error(error1));
24
25 endmodule
```

FIGURE 9-8. The Top Level of an Error Tree

- **Line 8**—Declare signals to whole errors from the `U0` instance and `U1` instance. Every block in this design has one `error` output.
- **Lines 10–14**—Create a flip-flop that stores a `1'b1` if one of the error signals goes high. The flip-flop stays high until we reset it.
- **Line 19**—Connect the `error` port from `U0` to `error0`.
- **Line 23**—Connect the `error` port from `U1` to `error1`.

You can also create the exact same type of top-level register in VHDL by creating a process that is sensitive to the clock and then clocks in the final error signal that was created by ORing all the lower signals together.

When we synthesize this we get a large OR tree of signals from each of the top-level modules. Inside each module is the assertion module that contains the checkers. Then the error signals from the checkers work their way up to the top.

Here is the synthesized error flip-flop as presented in Precision RTL:

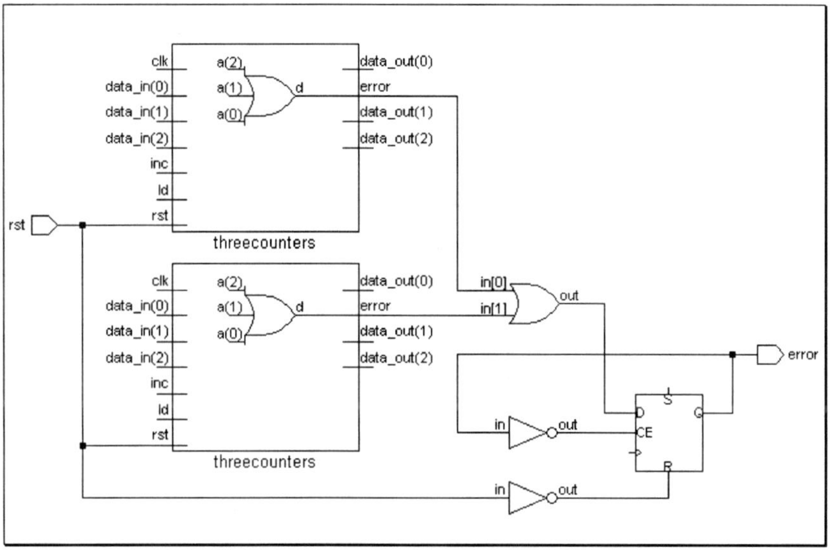

FIGURE 9-9. Error Flip-Flop in the Top Level

We now have a single output signal that tells us whether an assertion fired anywhere in our design.

9.5 Removing Assertion Hardware

Having assertions in your design makes it much easier to debug and lets you catch errors as soon as they happen. By creating an OR tree that brings the errors to the top of the design, you can create a visible signal with an LED or connect the signal to a logic analyzer to catch errors in your design.

Once you've debugged your design, you may want to leave the assertions in it so you can have a fault signal that can trigger in the field. The technician could tell you that the LED is lit, so you'd know that there is a bug in the FPGA.

On the other hand, you may be pressed for space. Perhaps you need to fit your design into the smallest FPGA possible and you can't afford to give all this logic to the debug effort. How can you synthesize without assertions? It's very easy.

The OVL allows you to turn off assertions on any synthesis run. When the assertions are turned off, all the checker modules disappear and then all the OR gates get optimized out of the design. Each language has its own way of turning off assertions.

9.5.1 Turning Off Assertions in Verilog

In Verilog you explicitly turn assertions on by defining the `'OVL_ASSERT_ON` macro. If you don't define this macro, then all the assertions disappear.

This schematic shows all the logic in our six-counter design, with the hierarchy flattened and `OVL_ASSERT_ON` defined. You can see, the design uses a significant amount of extra logic for assertions (at least from a graphical perspective). We'll compare this diagram to the one with no assertions in Figure 9-11 on page 115.

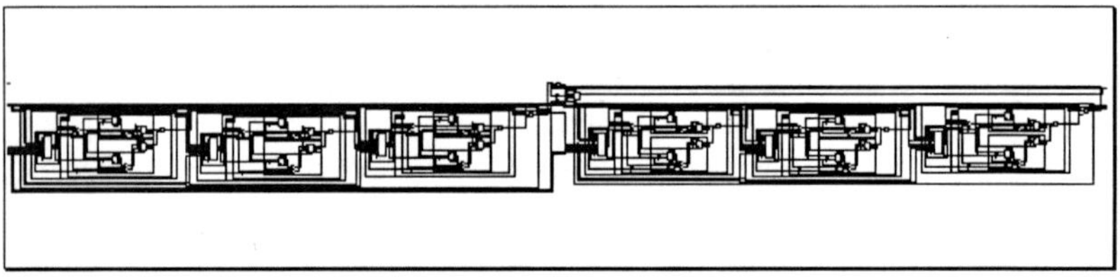

FIGURE 9-10. Assertion Logic in Six Counter Example

We explicitly asked Precision to create this logic by defining `OVL_ASSERT_ON`. To remove the assertion logic, we replace the line that defines `OVL_ASSERT_ON` in our synthesis script with the following line:

```
setup_design -defines {+define+OVL_VERILOG
        +define+OVL_SYNTHESIS}
```

We still define `OVL_VERILOG`, because we are not removing the OVL modules from the design. We are just not enabling the assertion logic. We could remove the `OVL_SYNTHESIS` macro, because it will not be used without assertions.

Once we remove OVL_ASSERT_ON we get the following, much cleaner-looking design:

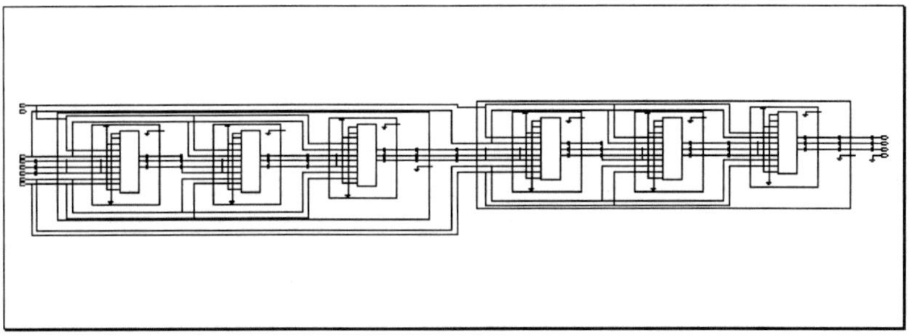

FIGURE 9-11. A Disappearing Assertion Tree

Not only did all the assertion modules disappear in this design, but even the top-level flip-flop was optimized away when it was clear that it would never change value. All the error signals are tied low.

9.5.2 Removing Assertions in VHDL

Unlike Verilog, VHDL requires that we say explicitly whether assertions are turned on or off. This is done in the control record. The file std_ovl.vhd defines a constant called OVL_ON and another one called OVL_OFF. We use these to turn features on and off in the VHDL OVL.

Normally we have assertions turned on and the control record looks like this:

```
package proj_pkg is
-- OVL configuration
  constant ovl_proj_controls : ovl_ctrl_record := (
-- generate statement controls
    xcheck_ctrl             =>    OVL_ON,
    implicit_xcheck_ctrl    =>    OVL_ON,
    init_msg_ctrl           =>    OVL_OFF,
    init_count_ctrl         =>    OVL_OFF,

    assert_ctrl             =>    OVL_ON,
```

FIGURE 9-12. VHDL Assertions Turned On

This control record gives us a three-bit counter with extra logic to implement `ovl_never`:

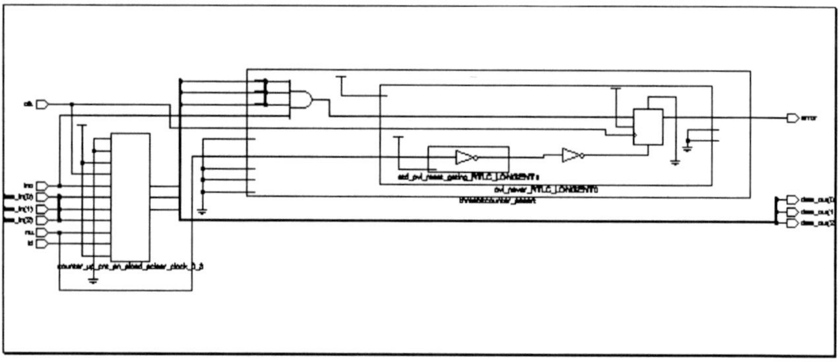

FIGURE 9-13. Three-bit Counter with Assertions On

You can see the error signal coming out of the assertion module and rising up to the top level.

We turn off the assertions like this:

```
4  package proj_pkg is
5  -- OVL configuration
6    constant ovl_proj_controls : ovl_ctrl_record := (
7  -- generate statement controls
8      xcheck_ctrl              =>     OVL_ON,
9      implicit_xcheck_ctrl     =>     OVL_ON,
10     init_msg_ctrl            =>     OVL_OFF,
11     init_count_ctrl          =>     OVL_OFF,
12
13     assert_ctrl              =>     OVL_OFF,  -- OVL_OFF disables assertions
14
```

FIGURE 9-14. VHDL Assertions Turned Off

We've used the `OVL_OFF` constant on line 13 to turn off assertions. Now VHDL will not generate the hardware needed to create the assertion.

Once we turn off assertions our counter looks like this:

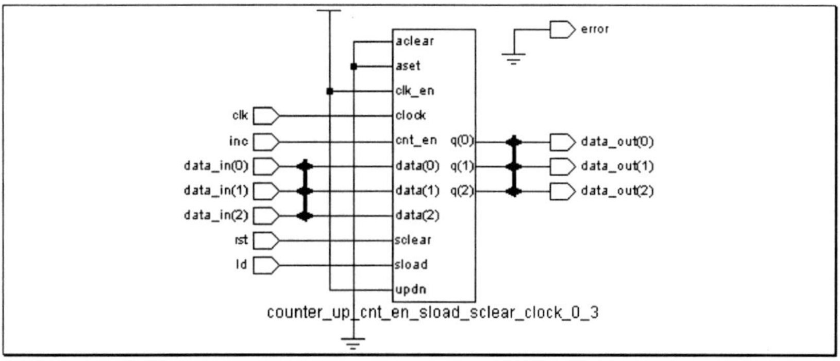

FIGURE 9-15. Assertion Block Optimized Away

There is no longer an `assertion` module; it has been optimized away. The OVL has tied the error signal low, because there are no assertions. This zeroed error signal travels up the OR tree and causes every OR gate in the design to disappear, along with the flip-flop at the top level.

9.6 Summary

In this chapter we learned how to take our OVL checkers out of the simulator and bring them into the lab. We learned which checkers are synthesizable, and we learned how the `fire` bus works on the output of all checkers. We discussed how to synthesize checkers in Verilog and VHDL.

Once we had a single checker synthesized, we saw how to create multiple checkers in one assertion module and combine their error signals with an OR gate. Then we brought the OR-ed error signals up through the hierarchy until we had a single error signal at the top of the design. This signal would latch if it became true, and this could be used to light an LED on the board or trigger a logic analyzer.

Finally, we saw how to turn off the assertions and remove all the assertion logic from our design.

We've now reached the end of Step 3, Assertions. Any design team would dramatically improve its productivity and reduce its risk in the lab if it simply implemented these three steps: turn on code coverage, create a test plan, add assertions.

But there is another level of productivity available. In these three steps we made no strides in how we wrote our tests. We had no techniques that helped us implement the channels and transactions we discussed in the test plan.

In our Step 4, Transactions, we'll address this weakness. We'll learn how to encapsulate the complexity in our test bench into modules and remove it from our tests. We'll learn how to create tests easily to test our RTL fully, and we'll see how we can use transactions to check quickly that our design is working the way we intended.

Until this point, we've been building a foundation with code coverage, test plans, and assertions. Now we'll build a robust and reusable test bench on that foundation in Step Four: Transactions.

10 Introduction to Transactions

IN THIS CHAPTER:

- *Transaction-Level Test Benches*
- *The TinyALU*
- *The TinyALU Transaction-Level Test Bench*

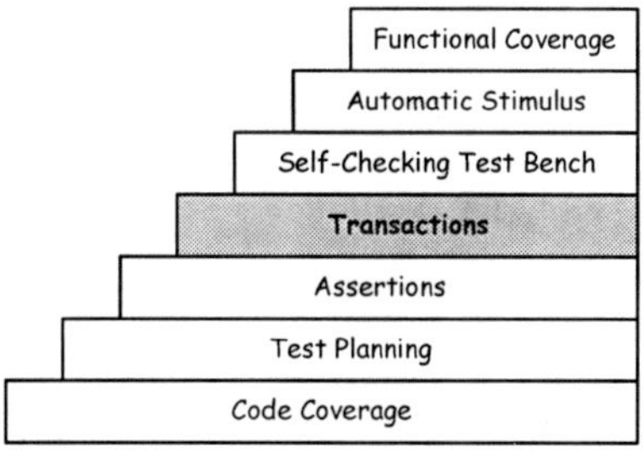

When we started climbing the staircase towards the ultimate test bench, we were running a few tests on our code and then heading out to the lab. Now we have a test plan, a measure of progress in code coverage, and a way to double our debugging productivity. We've come a long way.

The first three steps of the verification staircase form a powerful base for any team's progress. In fact, many teams could stop right here and live happily on their newfound productivity.

Other teams, however, will see that their bottleneck has simply moved from debugging the design to writing the tests. Their new test plans may contain dozens or even hundreds of variations on the ways in which the channels can interact, and writing a test for each of these may be impossible. These teams need a new style of test bench that makes it easy to write tests and check the results.

The *Transaction-Level Test Bench* answers their needs by making it easier to write tests and, in later steps, to have a test bench that writes its own tests and checks its own results.

10.1 Transaction-Level Test Benches

Digital design has always been a struggle against complexity. Billions of things happen every second in even the simplest design. The complexity gets worse when we look at designs that use multiple protocols or even embedded microprocessors.

Engineers respond to increasing complexity with higher levels of abstractions The first hardware designers worked with transistors. Then designers increased levels of abstraction by working with gates, MSI logic, high-level IP, and design languages.

All abstraction has the same goal. It allows us to design complex parts of our system once and then reuse the results. When we plunk an FPGA down on a board, we are reusing the complex design created by the FPGA engineers. We don't need to worry about how to design a chip by using the latest process because someone else has done that for us.

Abstraction also allows us to share design work across projects and companies. We can pass IP to other engineers who need to worry only about how to operate our block. They don't need to worry about how we implemented what's inside the blocks.

Transaction-level test benches do the same thing for simulation. They encapsulate the complexity of signal-level communication into modules. Then they allow you to write tests and check results without having to worry about the communication details. They also allow you to leverage test benches from other engineers, so that you can focus on test writing without having to think about signal pushing.

Once you don't have to worry about signals you can do the following with relative ease:

- Write tests that exercise all the combinations of transactions in your test plan.
- Reuse the transaction-level modules from your current design in any future design that uses the same channel.
- Create test benches that write their own stimuli and check their own results.

Using a transaction-level test bench gives you a tremendous boost in productivity and makes it easier to achieve the goal of 100% code coverage.

We examine transaction-level test benches from now until the end of the book, by using a simple design with a single input channel and a single output channel. We use a simple design so that we can focus on the test bench rather than the design functionality.[1]

10.2 The TinyALU

This design is called the TinyALU, a simple ALU written in VHDL. It accepts two eight-bit numbers (A and B) and produces a 16-bit result. Here is the top level of the TinyALU:

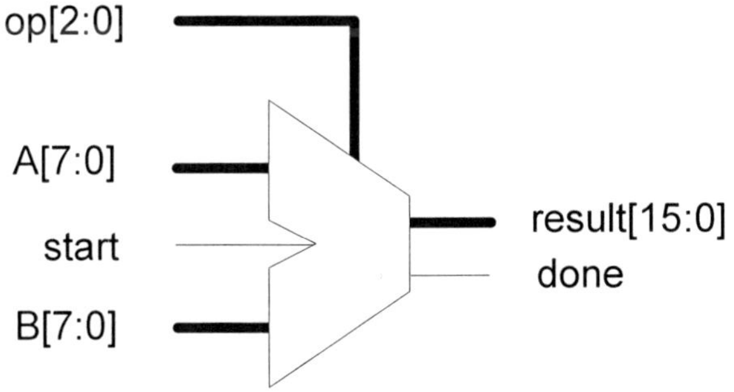

FIGURE 10-1. Top Level for the TinyALU

10.2.1 TinyALU Functional Spec

The TinyALU reads operands off the `A` and `B` buses and an operation off the `op` bus when the `start` signal is active. Then it delivers a result based on the operation on the `op` bus. Operations can take any number of cycles. The TinyALU raises the `done` signal when the operation is complete.

1. There are other, more advanced, designs available on www.fpgasimulation.com.

The TinyALU has five operations: NOP, ADD, AND, XOR, and MULT. The user encodes the operation on the three-bit `op` bus when requesting a calculation. Here are the encodings:

Operation	Opcode
NOP	000
ADD	001
AND	010
XOR	011
MULT	100
UNUSED	>100

FIGURE 10-2. Operation Coding for the TinyALU

Here is the waveform for the TinyALU:

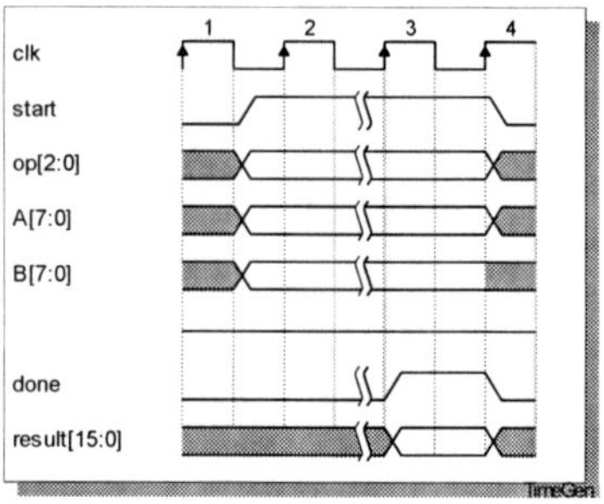

FIGURE 10-3. Waveform for the TinyALU[2]

The logic that initiates a TinyALU operation must keep the `start` signal high and the operands and operator stable until the TinyALU raises the `done` signal. The `done` signal stays high for only one clock. There is no `done` signal on a NOP operation, because there is no result. On a NOP, the requester lowers the start signal one cycle after raising it.

2. This diagram was created with TimeGen from Xfusion Software. TimeGen is a nifty little documentation tool. You can get it at http://www.xfusionsoftware.com/

This is the complete specification for the TinyALU. Now we create a transaction-level test bench that tests the TinyALU.

10.2.2 The TinyALU Test Plan

The TinyALU test plan follows the structure from Chapter 3. We are going to decompose the functionality of our design into channels and transactions.

There are two channels on the TinyALU:

TABLE 10-1. TinyALU Channels

Channel	Description
Request	Requests operations from the TinyALU.
Response	Returns the results.

Each channel on the TinyALU supports one kind of transaction.

TABLE 10-2. Transactions for TinyALU Channels

Request	Response
alu_operation(A, B, op)	alu_result(result)
A, B—8-bit values	result—A 16-bit number
op—One of the operations from Figure 10-2 on page 122	

We can define all of the TinyALU's functionality with these transactions, because the operation is very simple. The TinyALU creates an `alu_result` transaction for every `alu_operation` transaction unless the operation is a NOP.

The waveforms for the TinyALU transactions are in Figure 10-3 on page 122. The timing diagram defines the `alu_operation` transaction with the signals above the horizontal line, and the `alu_result` transaction with the signals below the horizontal line.

So far we've done nothing new. All the steps in this test plan are the same ones we'd follow to decompose any design and create tests for it. Now we are going to use this information in a more powerful way. We are going to define a test bench based on the channels and transactions we've just described.

10.3 TinyALU Transaction-Level Test Bench

In a transaction-level test bench, transactions move from being abstract concepts to being actual packages of data; they bring the channels of the test plan to life. Here is a generic view of a transaction-level test bench with an input and an output channel:

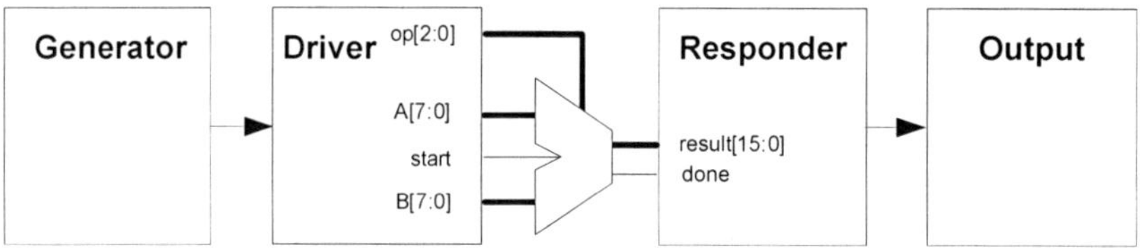

FIGURE 10-4. Transaction-Level Test Bench for the TinyALU

There are five modules in the top level of this test bench:

- **Generator**—This module creates transactions.
- **Driver**—The driver receives transactions from the generator and drives the data and control signals. This implements the request channel. In a DUT with several input channels, there would be several drivers.
- **TinyALU RTL**—The DUT doesn't know about transactions. It just sees the signals that connect it to the drivers and responders.
- **Responder**—The responder receives the data from the DUT and converts it back to transactions. This implements the response channel. In a DUT with several output channels, there would be several responders.
- **Output**– We do something with the transactions when we read them from the responder. In our example, we'll write them into a file for comparison with a expected results.

This is a simple design with only two channels. In a larger design, there is a driver for every input channel and a responder for every output channel.

We talked above about taking transactions from the generator module and putting transactions into the output module. We didn't discuss how transactions are represented. We talk about that in Chapter 11.

10.4 Summary

In this chapter, we discussed the power of abstraction and how we can create powerful test benches by abstracting away the details of driving signals. We saw that transaction-level test benches make it easier to write tests and to reuse parts of test benches in later projects.

We examined the TinyALU example. We'll use the TinyALU as an example for the rest of the book. We looked at the TinyALU functional specification and created the TinyALU test plan.

Once we had a test plan we saw, at a high level, how the transactions in the test plan can be converted into a test bench. The test bench takes transactions from a generator to create stimuli, and turns DUT results back into transactions.

We did not discuss the details of implementing transactions. We'll do that in the next chapter.

11 Creating Transactions

IN THIS CHAPTER:

- *Hello, SystemVerilog*
- *What are Objects?*
- *Defining Transactions*
- *Creating Transactions*

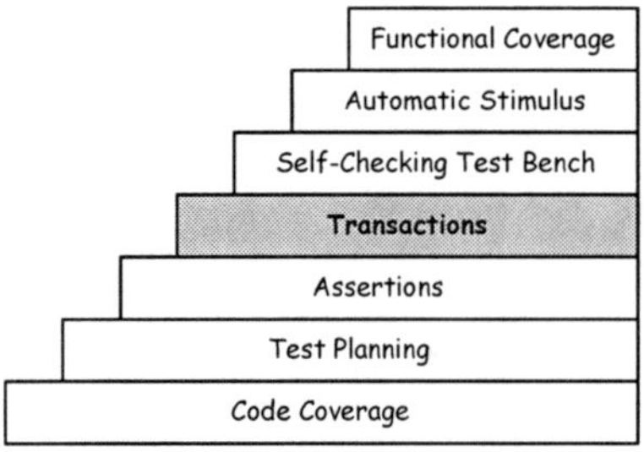

In the preceding chapter, we discussed the transaction-level test bench. This is a test bench where we encapsulate the complexity of dealing with a channel's signals into either a driver module or a responder module.

The driver module converts transactions into signals that drive the DUT, and the responder module converts signals back into transactions. What are these transactions? How do we create them and how do we move them between the modules? We'll learn how to create transactions in this chapter, and we'll learn how to move them between modules in the next chapter.

As part of our move to transaction-level test benches, it's time to introduce SystemVerilog.

11.1 Hello, SystemVerilog!

Until this point in the book, we've created our test benches in the same language as our designs. We'd test VHDL designs by writing VHDL, and we'd test Verilog by writing Verilog.

The problem with creating test benches with languages such as VHDL and Verilog is that these languages were designed for hardware creation, not test bench creation. So while they have powerful features for describing hardware, they lack features that support test bench creation.

Languages that have powerful test bench features are called high-level verification languages, or HVLs. There are several HVLs in the industry, including Vera and *e*. However, Vera and *e* are proprietary languages, each associated with a single simulation vendor. SystemVerilog is the first HVL based on an IEEE standard. As a result, there are several simulation companies that support SystemVerilog.

11.1.1 Why use SystemVerilog

As we move into transaction-level test benches, we will using SystemVerilog to test a VHDL device-under-test. SystemVerilog has three features that make it an excellent HVL:

- Data structures, such as objects, that make it easier to build advanced test benches.
- Built-in randomization to allow us to generate test stimulus.
- Coverage constructs that tell use whether we've tested all our functionality.

These features do not exist in RTL languages like Verilog and VHDL[1].

In addition to these features, we work with SystemVerilog because, as an industry standard, it has open libraries that support verification, such as an open library built by Mentor Graphics and Cadence Design Systems called the Open Verification Methodology (OVM).

Because of the power of SystemVerilog, we use it in all our transaction-level design examples. This combination of SystemVerilog with a VHDL DUT, allows VHDL-savvy engineers to continue using VHDL for their designs, while having access to the tools in SystemVerilog.

We are now ready to answer the first question of transaction-level test benches: "How do we represent transactions?" We do it with SystemVerilog objects.

1. Though VHDL does have several high-level programming constructs.

11.2 What are Objects?

Object-oriented programming is the subject of doctoral theses, conferences, books, and arguments in bars. It represents a dramatic leap forward in software design and we could discuss it forever.

But we won't.

We need to understand only a small part of object-oriented programming to write test benches, and so we'll focus on that part and leave the broader and more philosophical discussions to other books and other times.

Objects are like records in VHDL or structs in C: they contain references to a variables of different types. The difference between objects and records or structs is that objects can also contain tasks and functions that operate on the variables.

The data variables in an object are call *data members,* and the tasks and function in an object are called *methods*. Object definitions are called classes, and when we define an object we start with the keyword `class`.

We'll discuss objects by creating a simple class that represents a circle. The `circle` class stores the circle's radius and has methods that return the circle's diameter, area, and circumference.

Assuming we've defined a `circle` class, here's how we use it in a module. This example creates a new circle and prints the information about it:

```
25  module circle_mod;
26
27     circle c;
28
29     initial begin
30        c = new (2.5);
31        $display("Radius\t%4.2f", c.radius);
32        $display("Diameter\t%4.2f",c.diameter);
33        $display("Area\t\t%4.2f",c.area);
34        $display("Circumference\t%4.2f",c.circumference);
35     end
36
37  endmodule
```

FIGURE 11-1. Using the `circle` Class

- **Line 25**—This is a SystemVerilog module.

- **Line 27**—We declare a variable of type `circle`. This creates a variable called `c` that holds a pointer to a `circle` object.
- **Line 29**—An `initial` block is just like a VHDL `process` with a `wait` statement at the end of it.
- **Line 30**—We create a new circle with a radius of 2.5. Here's where SystemVerilog allocates new memory. We create an object of type `circle` with a radius of 2.5 and put the pointer to it into the variable `c`.

  We call pointers like this *handles*, a term we'll use from now on.
- **Line 31**—The variable `c.radius` is a data member of the `class` circle. We print its value here. The string `"4.2f"` is a C-like formatter that says to return a number four digits wide with two digits after the decimal point.
- **Lines 32–34**—There are three methods in this class: `diameter`, `area`, and `circumference`. They do the calculations to return these values.

We can see how the object we defined abstracts away the details of the circle. We just have to pass the object the radius of the circle. We previously put the smarts into the object definition to figure out all the statistics.

When we run this program we get the following results:

```
# Radius        2.50
# Diameter      5.00
# Area          19.63
# Circumference 15.71
```

FIGURE 11-2. Information About Our Circle

Now that we've used the `circle` class, let's see how it was defined. We define classes with the `class` keyword. Here's our `circle` definition:

```
1  class circle;
2
3    real radius;
4
5    function new(real r);
6       radius = r;
7    endfunction
8
9    function real diameter;
10     return radius * 2;
11   endfunction
12
13   function real area;
14     return 3.14159 * radius**2;
15   endfunction
16
17   function real circumference;
18     return 2 * 3.14159 * radius;
19   endfunction
20
21 endclass
```

FIGURE 11-3. Defining the `circle` Class

- **Line 1**—We define all classes with the `class` keyword. In this case we're defining a class called `circle`.
- **Line 3**—We define the data members. In this example we have just one data member: a real number called `radius`.
- **Lines 5–7**—The `new()` method is called the class's constructor. This one requires the radius as an argument. If you try to create a circle without providing a radius, you'll get an error.

  You need to provide a `new()` method only if your constructor has arguments.
- **Lines 9–11**—The `diameter` method doubles the `radius` data member and returns the value.
- **Lines 13–19**—We create two more methods that return the area and circumference of this object.
- **Line 21**—All class definitions end with the keyword `endclass`

We've now seen how to define a class in SystemVerilog and how to create a new object of that class. We've seen that objects can contain data of many types, just like a record in VHDL or C. The difference is that objects also contain tasks or functions that can operate on the data in the object.

11.3 Defining Transactions

Using SystemVerilog we can define transactions as classes and create transactions as objects. As we saw in Figure 10-4 on page 124, driver modules take transaction objects and translate them into signals, and responder modules take signals and translate them into transaction objects.

We create transactions by looking at transaction definitions in the test plan and creating object classes that match the data we need in each transaction. We define transactions in SystemVerilog packages and use the packages to pass the definitions around our test bench.

Like VHDL, SystemVerilog allows you to define types and declare variables in a common compilation unit called a *package*. Once you've created a package, you use it by importing it into the top of a module file. Here's an example of an `import` statement:

```
import tinyalu_pkg::*;
```

This statement says to take all the declarations and definitions in the `tinyalu_pkg` and make them available to the module in the file. We need to compile `tinyalu_pkg` into our library before we compile our module.

11.3.1 Common Methods in Transactions

In addition to the data members that we get from the test plan, each transaction needs four utility methods that are essential to creating a test bench. In anticipation of this, we define four methods in every transaction class. Then the methods will be ready for us to use when we need them. Here are the methods:

- **`new()`**—Defines all the data members.
- **`clone()`**—Returns a copy of the object. Cloning is essential when we start moving transactions through our test bench.
- **`convert2string()`**—Returns a string that contains the data in the object. We use the string to print messages that show us what's going on in our test bench.
- **`comp()`**—Compares this transaction to another of the same class. It returns a 1 if the two transactions are identical.

We see how to define these methods in our TinyALU example.

11.3.2 Defining Transactions for TinyALU Request Channel

In Table 10-2 on page 123 we saw that the TinyALU Request Channel had one transaction called `alu_operation`. Let's create that transaction as a SystemVerilog class.

The `alu_operation` has three pieces of data: the operation, the A-leg operand, and the B-leg operand. It also has the `new()`, `comp()`, `clone()`, and `covert2string()` methods.

We define our transaction in `tinyalu_pkg`. The package creates a new enumerated type to define the operations, and then we use that type in the `alu_operation` transaction:

```
1  package tinyalu_pkg;
2
3  typedef enum {add_op, and_op, xor_op, mul_op,no_op} operation_t;
4
5  class alu_operation;
6     rand logic[7:0] A,B;
7     rand operation_t op;
```

FIGURE 11-4. Defining the ALU Operation

- **Line 1**—Define a package called `tinyalu_pkg`.
- **Line 3**—Define an enumerated type called `operation_t` with one entry for each operation.
- **Line 5**—Define a class called `alu_operation`.
- **Line 6**—The class has two 8-bit data members called `A` and `B`.
- **Line 7**—The class has one `operation_t` data member called `op`.

The `op`, `A`, and `B` data members have an additional keyword in their declaration: `rand`. This keyword tells SystemVerilog that these data members are random variables.

All SystemVerilog objects define a method called `randomize()`. When you call the `randomize()` method on an object, SystemVerilog checks the object for random variables and randomizes them.

At this point the randomization is unconstrained. The variables can take on any possible value. In Step 6, Automatic Stimulus, we'll learn how to constrain random variables so they create useful tests. However, all values work for the TinyALU, so we'll use random variables to make it easy to create tests.

Now we need to define the five methods in the `alu_operation`:

```
   function new (
       logic[7:0] ia = 0,
       logic[7:0] ib = 0,
       operation_t iop = no_op );
      A  = ia;
      B  = ib;
      op = iop;
   endfunction

   function alu_operation clone();
      alu_operation c;
      c = new(A,B,op);
      return c;
   endfunction

   function bit comp (alu_operation t);
      return ((t.A == A) && (t.B == B) && (t.op = op));
   endfunction

   function string convert2string;
      string str;
      $sformat(str,"A: %2h  B:%2h   OP:%s", A,B,op);
      return str;
   endfunction

   function logic [2:0] op2bits;
      case (op)
         add_op : op2bits = 3'b001;
         and_op : op2bits = 3'b010;
         xor_op : op2bits = 3'b011;
         mul_op : op2bits = 3'b100;
         no_op  : op2bits = 3'b000;
      endcase // case(op)
   endfunction //op2bits()

endclass // alu_operation
```

FIGURE 11-5. Methods for `alu_operation`

- **`new(ia,ib,iop)`**—The `new()` method takes the values of `ia`, `ib`, and `iop` and loads them into the transaction. We've defined defaults for each of the arguments in case the user doesn't supply the data.
- **`clone()`**–The `clone()` method returns another transaction just like this one. It creates a new `alu_operation` called `c` and passes the constructor the current values for `A`, `B`, and `op`. Then it returns `c`.
- **`comp(t)`**—The `comp()` method compares this transaction to one that's been passed to the function. It returns a 1 if they are identical.
- **`convert2string()`**—This method creates a string with the data from the transaction and returns it. It displays `A` and `B` as hex values and `op` as a string that takes on the operation name.

- **op2bits()**—This method is unique to `alu_operation`. The driver uses this method to drive the correct opcode onto the operation bus. The numbers here match the ones in Figure 10-2 on page 122.

That's all there is to defining a transaction. We define the data and then we define the methods.

Here's another example: the complete definition of the `alu_result` transaction:

```
class alu_result;

    logic [15:0] result;

    function new( logic[15:0] iresult = 0);
       result=iresult;
    endfunction

    function alu_result clone();
       alu_result c;
       c = new(result);
       return c;
    endfunction

    function bit comp (alu_result r);
      return (r.result == result);
    endfunction

    function string convert2string;
       string str;
       $sformat(str,"Result: %h",result);
       return str;
    endfunction

endclass

endpackage
```

FIGURE 11-6. Definition of the `alu_result` Transaction

This closes out the `tinyalu_pkg`. We import this package into our modules to get access to these transaction definitions.

11.4 Creating Transactions

Now that we've defined our transactions as objects in a package file, it's time to create them. We create transactions in a module called a *tester*. The module references our package and has a variable that holds our transactions. We use that variable to call the `new()` method, and SystemVerilog allocates memory and creates new transactions for us.

Here's a simple tester for the TinyALU:

```
1  import tinyalu_pkg::*;
2
3  module top;
4
5     alu_operation op;
6
7     initial begin
8        op = new();
9        $display("Default:       ",op.convert2string());
10       op = new(8'h00,8'h01,add_op);
11       $display("Initialized:   ",op.convert2string());
12       op = new();
13       assert(op.randomize());
14       $display("Random:        ",op.convert2string());
15       op = new();
16       assert(op.randomize());
17       $display("Random again:  ",op.convert2string());
18    end
19 endmodule // generate_and_print
```

FIGURE 11-7. Tester for the TinyALU

- **Line 1**—We import `tinyalu_pkg` to get the `alu_operation` definition.
- **Line 3**—This is just a normal SystemVerilog module.
- **Line 5**—We declare a variable called `op` that holds handles to `alu_operation` objects.
- **Line 8**—We create a new transaction that uses default values.
- **Lines 9**—We display the object in `op` by using the `convert2string()` method that we defined in Figure 11-5 on page 134.
- **Lines 10–11**—We create a new transaction by passing it initial values. Then we display the transaction.

  This line raises an important question. What happened to the memory we allocated on line 8? We just overwrote the pointer to it.

  In C this would cause a memory leak—but not in SystemVerilog. SystemVerilog automatically deallocates memory if we are not pointing to it anymore. So the memory that was allocated on line 8 just disappears.
- **Lines 12–13**—We created a new transaction and then called the `randomize()` method on it. The `randomize()` method returns a 1 if it is successful, so we catch the return value with the `assert` statement.
- **Lines 14–17**—We output the randomized transaction. Then we create another one and output that.

Here's what it looks like when we run the module:

```
# Default:         A: 00  B:00    OP:no_op
# Initialized:     A: 00  B:01    OP:add_op
# Random:          A: c6  B:1d    OP:and_op
# Random again:    A: c4  B:bf    OP:no_op
```

FIGURE 11-8. Creating Transactions

In this example we created the transactions and printed their values. Now we're ready to start moving the transactions around the test bench.

11.5 Summary

SystemVerilog uses objects to define the data that goes into newly allocated memory. We saw that objects are like records. The difference between objects and records is that objects contain tasks and functions in addition to data fields. The data fields in an object are call *members,* and the tasks and functions are called *methods*.

We use objects to define the transactions in our transaction-level test bench. We saw that transactions can hold any value, and that they should contain the following methods: `new()`, `clone()`, `comp()`, and `convert2string()`.

We define our objects in a package, so the definitions are available to all the modules in our design. When we define variables in a package, we need to use `import` to bring the variables into a module.

After defining the `alu_operation` transactions, we generated a few and printed their values to the screen. We saw that we could either use default values for the transaction, passing it the values we want, or let SystemVerilog randomize it.

Transaction-level test benches pass transactions between modules. We've seen how to create the transactions; in the next chapter we'll see how to move them between modules.

12 Threads and FIFOs

IN THIS CHAPTER:

- *Creating Multiple Threads*
- *The Open Verification Methodology*
- *Synchronizing Threads*
- *Transaction Level Modules*

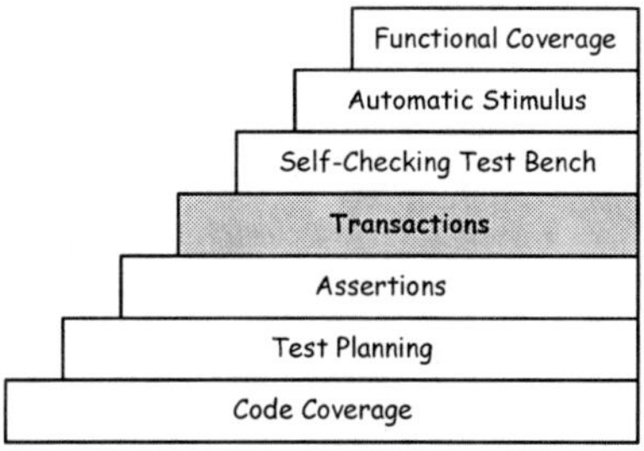

In our previous chapter we saw how to create transactions for our transaction-level test bench. We generated some transactions and printed them to the screen. Now it's time to start moving those transactions around the test bench.

Recall that we are working to create a test bench that looks like this:

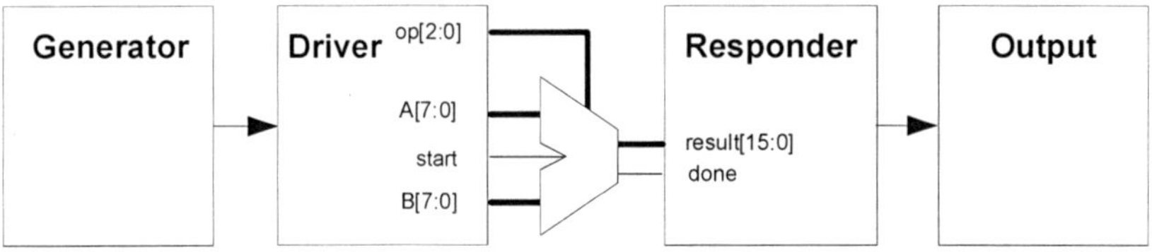

FIGURE 12-1. Transaction-Level Test Bench

The generator module makes transaction objects and sends them to the driver module. The driver module converts the transactions to signals to stimulate the DUT (the TinyALU in this case). The responder module reads signals from the DUT and converts them to transactions. Finally, the output module gets the transactions from the responder and does something with them, such as write them to a file.

The modules in this test bench work like the blocks in an RTL design. Each module runs one or more threads in parallel. We model hardware with this multi-threaded behavior.

In RTL design our VHDL `process` blocks or Verilog `always` blocks launch themselves in their own threads. We don't have to do anything to launch multi-threaded behavior. These blocks loop forever, executing their functionality.

VHDL and Verilog control their infinitely looping blocks with sensitivity lists. These language constructs sit at the top of the loop and make it suspend until a certain event happens. This ability to suspend threads is critical to simulating hardware.

In transaction-level test benches, we launch the threads ourselves. We also manage the synchronization between the threads differently than we do in RTL. In RTL we rely on the sensitivity lists to synchronize behavior between our blocks. In transaction-level test benches we use other mechanisms for synchronization.

This chapter explains how to create multiple threads and how to make them communicate with each other.

12.1 Creating Multiple Threads

It's easy to create multiple threads in VHDL and Verilog. You just write a `process` block or an `always` block and then start simulating. The thread creation process for transaction-level test benches is similar to the RTL process, except that we write tasks and launch the threads manually.

SystemVerilog tasks are the same thing as VHDL procedures. They are subprograms that encapsulate behavior. We launch these tasks in individual threads when we create a transaction-level module.

We launch threads in SystemVerilog (and in Verilog) with a language construct called `fork...join_none`. We use a `fork` statement to tell SystemVerilog that all the statements in the block should get their own thread. Use use the `join_none` statement to close the list of threads.

In this section we'll demonstrate multi-threaded programming with two tasks in a single module. Then we'll look at how to synchronize the tasks, and finally we'll look at how to put the tasks into multiple modules and get the modules to talk together. You'll see `fork...join_none` in action.

Here's an example of a simple multi-threaded module. It demonstrates the steps of creating tasks and launching them in parallel. The module has two tasks, `producer` and `consumer`. The producer creates integers, and the consumer reads them. They both share the `data` variable to implement this communication:

```
16  module first_threads ;
17
18  int data;
19
20  task producer;
21    for (data = 0; data<=5; data++) begin
22       $display("Put Data: %d",data);
23    end
24  endtask
25
26  task consumer;
27    for (int i = 0; i <= 5; i++) begin
28      $display("Get Data: %d",data);
29    end
30  endtask
31
32  initial
33    fork
34      producer;
35      consumer;
36    join_none
37
38  endmodule
```

FIGURE 12-2. The `first_threads` Module

- **Lines 20–24**—The `task...endtask` keywords define a SystemVerilog task. The `producer` task increments the `data` variable until `data` is greater than 5. It prints the `data` variable on each loop. The `data` variable is global, so the `consumer` task can read it.
- **Lines 26–30**—This is a related task called `consumer`. It loops six times and prints the `data` variable to the screen.
- **Line 32**—A SystemVerilog `initial` block is like a VHDL `process` block with a `wait` statement at the end. It launches at time zero and runs once.
- **Line 33**—The `fork...join_none` block tells SystemVerilog to create a separate thread for each statement in it.
- **Lines 34–35**—We make one thread to run the `producer` task and a separate thread to run the `consumer` task. These launch as soon as they can.
- **Line 36**—The `join_none` line terminates the `fork` block without waiting for the threads to terminate. We've now launched two threads.

Looking at these two threads, we might expect to see the `producer` thread writing numbers into `data` and the `consumer` thread reading them. Unfortunately, we don't see that. We see the following instead:

```
# Put Data:        0
# Put Data:        1
# Put Data:        2
# Put Data:        3
# Put Data:        4
# Put Data:        5
# Get Data:        6
# Get Data:        6
# Get Data:        6
# Get Data:        6
# Get Data:        6
# Get Data:        6
```

FIGURE 12-3. Result of `first_thread` Module

Well, this obviously didn't work. The `producer` counted from zero to five and then the `consumer` read all sixes. What's up with that?

The problem is that the `consumer` didn't get to run until the `producer` was finished. This example demonstrated a key fact about programming:

Multi-threaded programming is an illusion.

Even though we forked off `producer` and `consumer`, they really aren't running in parallel. There's only one simulator, and both tasks use it until they are forced to suspend. In this case, the `producer` got the simulator first, so it ran through all its code. When the `producer` was done, the `consumer` had its chance. Because the `producer` had completed its `for` loop, the `consumer` saw only the last value of `data`, which was `6`.

SystemVerilog threads hold onto the simulator until they are forced to give it up. RTL threads use a sensitivity list to suspend threads. Because tasks don't have sensitivity lists, we need some extra help to synchronize our tasks. This help comes from the Open Verification Methodology (OVM).

12.2 The Open Verification Methodology (OVM)

When we added assertions to our test bench, we took advantage of the work of others and used the Open Verification Library (OVL) to instantiate checkers into our design.

Now that we've added transactions to our test bench, it's time to use another library that starts with the words "Open Verification." In this case, however, we are talking about the Open Verification Methodology, or OVM. The OVM is not related to the OVL, except that both are free and both provide intellectual property (IP) that makes our lives easier.

The OVM delivers a set of verification objects that run on any simulator that supports SystemVerilog. It is the result of collaboration between Cadence Design Systems and Mentor Graphics.

The objects in the OVM are powerful, and learning how to use them all would take another book. Fortunately, we need to take only one class definition and one object out of the library to get our job done:

- **tlm_fifo**–This *class* synchronizes data between modules.
- **ovm_top**—This *object* controls how messages get written to the screen. The OVM automatically creates this object before we start simulating.

The TLM in tlm_fifo stands for transaction-level modeling. We will use the acronym TLM from now on when we talk about transaction-level test benches.

12.2.1 Downloading and Compiling the OVM

You can download the OVM from the web site www.ovmworld.com. This gives you a zip or tar file that you unzip or untar to get a directory called ovm-#.#, where the # symbols are the digits of the version number of the OVM.

Once you have the directory in place, you compile it with the following command:

```
<compile command> +incdir+ovm-#.#/src ovm-#.#/src/ovm_pkg.sv
```

where <compile command> is whichever Verilog compilation command works with your simulator. This creates a package called ovm_pkg. All we need to do now is import that package in order to have access to all the OVM objects.

12.3 Synchronizing Threads

The tlm_fifo demonstrates the beauty of abstraction. The OVM developers spent a lot of time designing, writing, and debugging this object so that it would allow threads to talk to each other. We don't need to worry about how it was written. We just need to follow three steps:

1. Declare a variable that holds the tlm_fifo.
2. Create a new tlm_fifo.
3. Use the new tlm_fifo.

The tlm_fifo transfers data between threads. It has several methods to allow threads to put data into the FIFO and get it out. We discuss two of these methods in this chapter:

- **put()**—The put() method places data into the tlm_fifo. If the tlm_fifo already contains data, then the put() method suspends. The method has a single argument, the data you're putting into the tlm_fifo.
- **get()**–The get() method removes a piece of data from the tlm_fifo. If the tlm_fifo is empty, then the get() method suspends. The method has a single argument, the variable that receives the new datum.

Let's fix the producer and consumer in our first thread example by using the tlm_fifo. We wanted to see the consumer print each value created by the producer, but instead we saw the producer write all its values at once, while the consumer got to read only one value. We'll fix this by passing the data through the tlm_fifo.

We've rewritten the producer/consumer example and called the new module good_threads. The biggest difference between this code and the first_threads code is that we used a tlm_fifo called fifo to synchronize the communication between the threads.

```
1  import ovm_pkg::*;
2
3  module good_threads;
4
5  tlm_fifo #(int) fifo;
6
7  task producer;
8    int data;
9    for (data = 0; data<=5; data++) begin
10      fifo.put(data);
11      $display("Put Data: %d",data);
12   end
13 endtask // producer
14
15 task consumer;
16   int data;
17   forever begin
18     fifo.get(data);
19     $display("Get Data: %d",data);
20   end
21 endtask // consumer
22
23 initial begin
24   fifo = new("fifo");
25   fork
26     producer;
27     consumer;
28   join_none
29  end
30
31 endmodule
```

FIGURE 12-4. Correctly Synchronized Threads

- **Line 1**—We import the `ovm_pkg` package. Now we have access to all the OVM classes and objects.
- **Line 5**—We declare `fifo` to be a handle to a `tlm_fifo` object that can hold data of type `int`. Whenever you use a `tlm_fifo`, you need to tell it what kind of data passes through it.
- **Lines 7–13**—The producer's behavior is slightly different now. It still loops data through the numbers from 0 to 5, but now it puts `data` into `fifo` on line 10 by using the `put()` method.
- **Lines 15–21**—Now `consumer` can also access the global `fifo` variable. On line 18 `consumer` gets an integer from `fifo` with the `get()` method.
- **Lines 23**—This initial block starts things up.
- Line 24—We must create the new `fifo` object before we launch the threads. If we don't create `fifo` before launching the threads we will get run time errors as the threads try to use the object.

  Notice that we pass the string `fifo` to the `tlm_fifo` constructor. We do this because the `tlm_fifo` expects us to provide its name when we call it. By convention we pass a string that is the same as the variable name. We'll see later that other types of objects expect other types of information in their constructors.

- **Lines 25–28**—These are the same `fork` and `join` statements as in the `first_thread` module. They launch the threads that run the `producer` and `consumer` code.

This code delivers the behavior we want. The `producer` creates the numbers from 0 to 5, and the `consumer` reads the numbers from 0 to 5. Here is our simulation running properly:

```
# Put Data:          0
# Get Data:          0
# Put Data:          1
# Get Data:          1
# Put Data:          2
# Get Data:          2
# Put Data:          3
# Get Data:          3
# Put Data:          4
# Get Data:          4
# Put Data:          5
# Get Data:          5
```

FIGURE 12-5. Properly Synchronized Producer and Consumer

The `good_thread` module worked because we synchronized the communication between the threads with a `tlm_fifo` object.

The nice thing about the `put()` and `get()` methods is that we can write code that ignores the fact that it's multi-threaded and being synchronized. We simply call the methods where it makes sense to call them and SystemVerilog does the rest.

The `tlm_fifo` has created the exact kind of communication we need to implement transaction-level modeling. Now we need to use these techniques to get modules to talk to each other.

12.4 Transaction-Level Modules

Now that we know how to make threads communicate, we can move to creating modules that communicate at the transaction-level. We'll use modules to separate the different pieces of our test bench as we saw in Figure 12-1 on page 139.

Transaction-level modules communicate in much the same way as RTL-level modules. With RTL modules, you create wires at the top level of the design and connect them to the module port lists. Then the modules drive the wires to communicate.

With transaction-level modules you create `tlm_fifos` at the top level and connect them to the module port lists. Then the modules put transactions into the `tlm_fifos` in order to communicate. Below, we implement basic module communication by means of `tlm_fifos` by implementing our `producer/consumer` example as modules.

This module demonstrates the basic features of all transaction-level test benches:

```
1  import ovm_pkg::*;
2
3  module top ;
4    tlm_fifo #(int) fifo;
5
6    producer prod(.data_f(fifo));
7    consumer cons(.data_f(fifo));
8
9    initial begin
10     fifo = new("fifo");
11     fork
12        prod.run();
13        cons.run();
14     join_none
15   end
16 endmodule
```

FIGURE 12-6. Producer/Consumer Top Level

- **Line 1**—We import the OVM class library to get access to the `tlm_fifo` definition.
- **Line 4**—We declare `fifo` to be a type `tlm_fifo` accepting `int` data types.
- **Line's 5–6**—We instantiate a `producer` module and a `consumer` module and attach `fifo` to their `data_f` ports. As we'll see, these modules are defined to accept a `tlm_fifo` on their port.
- **Line 10**—Before we can launch the threads that drive this design, we need to create the `tlm_fifo` that the modules use to communicate. If we forget this step, we'll get a runtime error.
- **Lines 11–14**—By convention, our TLM modules have a task called `run`. We'll use the `fork...join_none` construct to launch all these `run` tasks in separate threads.

That is all there is to creating the top level of a TLM test bench. The `producer` and `consumer` modules do all the work.

The `producer` module contains the same functionality as the `producer()` task we discussed in Section 12.3. It creates integers and puts them into the `tlm_fifo`:

```
1  import ovm_pkg::*;
2
3  module producer (ref tlm_fifo #(int) data_f);
4
5    task run;
6      int data;
7      string msg;
8      for (data = 0; data<=5; data++) begin
9         data_f.put(data);
10        $display("Put Data: %0d",data);
11     end
12   endtask
13 endmodule
```

FIGURE 12-7. Producer Module Source Code

- **Line 3**—This module accepts a reference to a `tlm_fifo` on its `data_f` port. This is why the top level was able to connect the modules by using a `tlm_fifo`.
- **Line 5**—We have renamed the `producer` task from Section 12.3 to "`run`." This is a convention in TLM test benches. By naming all the primary tasks `run`, we make it easy to launch them from the top level.
- **Line 9**—We put the value of data into `data_f`, which points to the top-level `tlm_fifo`. The `put()` method suspends if the `tlm_fifo` is full.

This module creates numbers from 0 to 5 and passes them to the `consumer` module, which has the same sorts of changes as the `producer`:

```
import ovm_pkg::*;

module consumer (ref tlm_fifo #(int) data_f);

  task run;
    int data;
    forever begin
      data_f.get(data);
      $display("Get Data: %0d",data);
    end
  endtask

endmodule
```

FIGURE 12-8. Consumer Module Source Code

We've created a `run` task that takes data out of `data_f` and prints it to the screen. The `get()` method suspends if there is no data in the `tlm_fifo`.

We now have two modules that communicate entirely through `tlm_fifos`. The top level makes the connection and launches the threads.

When we run this design we get the following:

```
# Put Data: 0
# Get Data: 0
# Put Data: 1
# Get Data: 1
# Put Data: 2
# Get Data: 2
# Put Data: 3
# Get Data: 3
# Put Data: 4
# Get Data: 4
# Put Data: 5
# Get Data: 5
```

FIGURE 12-9. Multi-Threaded Modules in Action

12.5 Summary

In this chapter we learned how to create threads with the `fork...join_none` construct by launching SystemVerilog tasks in their own threads. We saw that threads don't really run in parallel and that thread communication requires special attention.

We learned about the Open Verification Methodology (OVM), a library of SystemVerilog classes and objects that help us create test benches. While this library is enormous, we are going to use only the `tlm_fifo` class and the `ovm_top` object from the library.

We synchronized our threads by using the `tlm_fifo` objects from the OVM. Then we saw how to use `tlm_fifos` to connect threads from different modules. We saw that SystemVerilog modules accept a reference to a `tlm_fifo` object on their port list, and we synchronized two modules.

We're now ready to create a transaction-level test bench. First, however, we are going to stop to talk about the `ovm_top` object and how it help us report data from our simulation. Once we understand the reporting functions, we'll be ready to build and debug transaction-level test benches.

13 OVM Reporting Tools

IN THIS CHAPTER:

- *Basic OVM Reporting Methods*
- *Advanced Reporting Control*

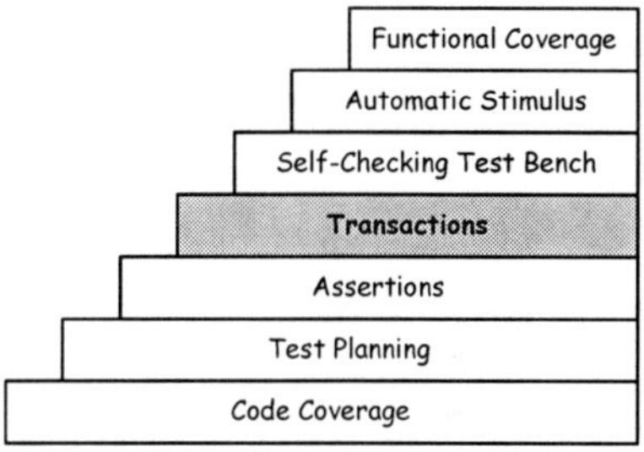

Writing tests at the transaction level allows us to create millions of transactions with only a few lines of code. We'll see soon how to write tests that can literally run forever.

However, with great stimulus comes great output. We can generate millions of messages to the screen, but generating this many messages has two problems:

1. It dramatically slows down testing. Every time we write a message our simulation stops.
2. It floods us with information that we need to filter. We can't find the word "ERROR" if it's hidden among ten thousand instances of the word "NOTE." So we have to write scripts to search our output.

The solution to this problem is simple. We filter our output—easier said than done. If we are using simple `$display` statements all over the place, then we need either to (a) encapsulate them with conditional compilation, or (b) provide `if` statements that control whether they print. We also have challenges telling `$display` statements where to print. Consequently, we really don't want to use `$display` as a reporting system.

Fortunately, the OVM solves this problem for us with an object called `ovm_top`. This object has the OVM reporting system built into it.

The OVM reporting system lets us do the following:

1. Reduce the number of messages being written.
2. Create different reporting behaviors for different modules.
3. Write messages to multiple log files without printing them to the screen.
4. Stop our simulation if we have too many errors.

OVM reporting gives us the power to manage all of this, and it all starts with the four basic reporting methods.

13.1 Basic OVM Reporting Methods

OVM reporting has four basic reporting methods. We always use one of these methods to print messages to the screen or to a file. Here are the four methods:

- `ovm_top.ovm_report_info`
- `ovm_top.ovm_report_warning`
- `ovm_top.ovm_report_error`
- `ovm_top.ovm_report_fatal`

All these methods have five arguments:

```
ovm_top.ovm_report_xxxxxxx(
string <id>, string <message>,
int <verbosity>,
string <filename>,int <line>)
```

The first two arguments are mandatory:

- **id**—This is a string that the system uses to sort and identify the report line. As we'll see below, you can use this ID to control the reporting system. The OVM counts the number of messages associated with each ID.
- **message**—This is the string you want to write to the file. You'll include things like `Danger, Will Robinson!` or whatever you deem appropriate for your messages.
- **verbosity**—You set a verbosity number for each message and then a global verbosity when you run the simulation. The OVM prints only messages whose verbosity is below the global verbosity number.
- **filename**—If you'd like to print the filename so you can find your messages, you can add it here.

- **line**—This is the line number in the file where the message is written.

These are the four basic reporting methods in action:

```
1  import ovm_pkg::*;
2
3  module top;
4   string m;
5   initial begin
6     $sformat(m,"%m");
7     ovm_top.set_report_verbosity_level(200);
8     ovm_top.ovm_report_info    (m,"Just Info",     300, `__FILE__,`__LINE__);
9     ovm_top.ovm_report_warning(m,"A Warning!",    200, `__FILE__,`__LINE__);
10    ovm_top.ovm_report_error   (m,"Normal Error", 100, `__FILE__,`__LINE__);
11    ovm_top.ovm_report_fatal   (m,"Fatal Error" ,   0, `__FILE__,`__LINE__);
12  end
13 endmodule
```

FIGURE 13-1. Using the Four Default Reporting Methods

- **Line 1**—We import `ovm_pkg` to get access to the `ovm_top` object.
- **Line 4**—We are going to store the module path in a string called `m`.
- **Line 6**—The `$sformat` system call uses the `%m` formatting string to store the current module path into `m`.
- **Line 7**—Whenever we print a message, we supply a verbosity number. The `ovm_top` object prints only messages with verbosity numbers below the level we set here.
- **Lines 8–11**—We print one message with each of the severity levels. Each message also gets a different verbosity level (0 to 300).

  We use the `` `__FILE__ `` (two underscores before and after) and `` `__LINE__ `` macros to pass the message the source file number and line number.The `` `__FILE__ `` and `` `__LINE__ `` macros in these calls tell you where the message originated in the source code. If you have multiple instances of the same module, the `m` string gives you the path to the message call.

As we'll see in Figure 13-2, not all of these calls get executed. We set the verbosity to 200, but call `ovm_report_info()` with a verbosity of 300. The `ovm_top` object filters out the messages with a verbosity higher than 200.

When we run this program we get the following:

```
29  # ---------------------------------------------------------------
30  # OVM-2.0
31  # (C) 2007-2008 Mentor Graphics Corporation
32  # (C) 2007-2008 Cadence Design Systems, Inc.
33  # ---------------------------------------------------------------
34  # OVM_WARNING top.sv(9) @ 0 [top] A Warning!
35  # OVM_ERROR top.sv(10) @ 0 [top] Normal Error
36  # OVM_FATAL top.sv(11) @ 0 [top] Fatal Error
37  #
38  # --- OVM Report Summary ---
39  #
40  # ** Report counts by severity
41  # OVM_INFO :    0
42  # OVM_WARNING :    1
43  # OVM_ERROR :    1
44  # OVM_FATAL :    1
45  # ** Report counts by id
46  # [top                    ]     3
47  # ** Note: $finish    : ../../ovm/src/base/ovm_report_object.svh(149)
48  #    Time: 0 ns  Iteration: 0  Instance: /ovm_pkg::ovm_report_object::die
```

FIGURE 13-2. Results of the Four Reporting Methods

- **Lines 29–33**—The OVM starts every simulation run with this header. It gives us the version number and the names of the contributing companies.
- **Lines 34–35**—These three calls correspond to the four function calls in Figure 13-1 on page 153. Why only three lines when we had four calls? Because we set the verbosity level to 200 on line 7 of the source code. The call to `ovm_top.ovm_report_info()` had a verbosity of 300, so it was filtered out.
- **Lines 38–48**—When we call `ovm_top.die(),` it waits to the end of the simulation cycle, prints out a summary of the number of times we called each warning level and the number of times we used each string ID, and then ends the simulation.

  But, you say, we didn't call `ovm_top.die()`. Actually we did, because `ovm_top.ovm_report_fatal()` calls `ovm_top.die()` by default.

The verbosity functionality in these basic reporting methods makes them much more valuable than `$display`. Normally, when we debug we put debug print statements in our code and then have to remove them later or comment them out. By using `ovm_top`, we can give these messages a high verbosity, say 500, and then set verbosity to 200 (or even 0) if we don't want to see extra messages.

Notice that the summary gave us two counts. It counted the number of times we used each severity (lines 40–44), and it counted the number of times it saw each ID (lines 45–46.)

These variables, severity and ID, are important because they are the basis for the advanced forms of reporting that we'll discuss in the rest of this chapter.

13.2 Advanced Reporting Control

We can create a usable test bench with the four basic message calls and the ability to set the verbosity level on a simulation. However, we will want to exert more control than these simple calls allow.

We control our messages by using two arguments common to all message routines:

- **The ID string**—This is the first argument to all reporting methods. In our example we used the module path as the ID string. We can tell `ovm_top` to do different things for different ID strings.
- **The severity**—The severity is inherent to our reporting methods. There are four severities: information, warning, error, and fatal. These correspond to `ovm_report_info()`, `ovm_report_warning()`, `ovm_report_error()`, and `ovm_report_fatal()`.[1]

You can control two things by using the ID string or the severity:

- **Actions**—You can control which actions happen when you call the reporting methods. We'll discuss actions in the next section.
- **Log Files**– The `ovm_top` object can log messages to a file. Each ID string, severity, or combination of these can have its own file name.

These control knobs allow you to say things such as, "Log all messages from the driver to the `driver.log`" or "Ignore all warnings." You can also say things such as, "Quit the simulation after five errors."

First we'll look at the available actions, and then we'll look at managing files. Finally, we'll put the two together in the `ovm_top` methods that control reporting.

1. The rest of this chapter assumes the "`ovm_top.`" part of all reporting calls. So we'll write `ovm_top.ovm_report_info()` as `ovm_report_info()`.

13.2.1 Actions

There are five actions available for all messages in the OVM. These are defined in an enumeration called `ovm_action_type` in the `ovm_pkg` package. This enumeration has seven values:

- `OVM_NO_ACTION`—As the name implies, this action does nothing.
- `OVM_DISPLAY`—Prints the message in the call to the screen. By default, all the OVM message routines do the `DISPLAY` action.
- `OVM_LOG`—Writes the message to a log file. By default, `LOG` is turned off; if you want to use it, you need to open a file and pass its handle to `ovm_top`. You can open multiple files and have the messaging system write to them based on the severity, message ID, or both.
- `OVM_COUNT`—Increments an error counter inside `ovm_top`. You can set `ovm_top` to shut down if there are too many errors. *The OVM counts errors only if you've set a maximum quit count in* `ovm_top`.
- `OVM_EXIT`—This action causes `ovm_top` to shut down the simulation in an orderly fashion. It waits for the end of the current simulation time step, closes all files, prints a summary of the simulation, and calls `$finish`. This action is set by default in `ovm_report_fatal()`.
- `OVM_CALL_HOOK`—This action calls a routine that you specify. We do not discuss this action in detail here.
- `OVM_STOP`—This is the equivalent of a breakpoint. It stops the simulation as soon as you call the reporting message that you've associated with this action. Then you can debug from there.

We can attach these actions to any severity level or message ID. We can also combine them to create multiple actions in one call to the reporting system.

By default, the `ovm_top` object connects these actions to severity levels, which work the same way for all message IDs. We remember that there are four severity levels: information, warning, error, and fatal. Here's how these are connected to actions:

TABLE 13-1. Default Actions for Severity Levels

Severity Level	Actions
INFORMATION	`OVM_DISPLAY`
WARNING	`OVM_DISPLAY`
ERROR	`OVM_DISPLAY, OVM_COUNT`
FATAL	`OVM_DISPLAY, OVM_EXIT`

Notice that the error and fatal severities execute two actions. This is because `ovm_top` combined the actions for these severities. We can combine actions ourselves to get the OVM to do what we want when we call a reporting method.

To create our own actions, we need to understand how the `ovm_pkg` defines the `ovm_action_type`. Here is the source code that defines the type:

```
typedef enum ovm_action
{
  OVM_NO_ACTION = 6'b000000,
  OVM_DISPLAY   = 6'b000001,
  OVM_LOG       = 6'b000010,
  OVM_COUNT     = 6'b000100,
  OVM_EXIT      = 6'b001000,
  OVM_CALL_HOOK = 6'b010000,
  OVM_STOP      = 6'b100000
} ovm_action_type;
```

FIGURE 13-3. Defining `ovm_action` in OVM Package

The OVM developers defined each action in the `ovm_action_type` as a six-bit value with one bit set. This allows us to create multiple actions by ORing the enumerated types together to create a new action.

For example, we can create an action that displays, counts, and logs:

```
ovm_action_type dcl; // display, count, and log

dcl = OVM_DISPLAY | OVM_COUNT | OVM_LOG;
```

This code does a bitwise OR on `DISPLAY`, `COUNT`, and `LOG`, and gives the following value: `6'b000111`. The bits in this string indicate which actions execute.

We'll use these values below when we change OVM reporting actions according to the severity or the message ID.

13.2.2 Log Files

Verifying a design by writing messages to the screen is fraught with peril. I remember a scene in the movie *The Andromeda Strain* where a doctor missed a key warning message on a computer screen because its rate of flashing just happened to set off her epilepsy. The world was almost destroyed by her failure to see the message, which would not have happened if she had logged her messages to a file.

We'll avoid her fate by logging all our messages to a file and then inspecting them later. In fact, we should not bother writing to the screen except for status messages and for errors.

In order to log our messages to a file, we need to open the file (or files) and pass the file handles to `ovm_top` so it knows where to put our information. We open files in SystemVerilog with the `$fopen()` system task.

The `$fopen()` system task can open files for reading or writing. It can append to existing files. It can read or write text or binary information. It is a powerful system call, but we are going to ignore most of that power in the name of simplicity. We are going to use `$fopen()` only in the following way:

```
integer file_handle = $fopen("filename");
```

When we call `$fopen()` with one string argument, it opens the file for writing and returns an integer with one bit set. This integer is called a *multichannel descriptor*, or MCD.

The MCD has one bit set for every file that you've opened for writing. This means that you can write to as many as 32 files simultaneously by ORing the MCDs together. For example, let's make three MCDs:

```
1  module top;
2
3  initial
4     begin
5        integer filea, fileb, fileab;
6        filea = $fopen("allmsgs.txt");
7        fileb = $fopen("somemsgs.txt");
8        fileab = filea | fileb;
9        $fdisplay(filea,  "This goes to A");
10       $fdisplay(fileb,  "This goes to B");
11       $fdisplay(fileab, "This goes to both");
12       $fclose(filea);$fclose(fileb);$fclose(fileab);
13    end
14 endmodule
```

FIGURE 13-4. Writing to Files with MCDs

- **Line 5**—Declare three integer variables to hold the MCDs.
- **Line 6**—Open the file named `allmsgs.txt`.

- **Line 7**—Open the file named `somemsgs.txt`. This puts an MCD with a different bit set into the `fileb` variable.
- **Line 8**—Create a new MCD by ORing together the `filea` and `fileb` MCD variables.
- **Line 9**—Write only to the `filea` file with `$fdisplay`.
- **Line 10**—Write only to the `fileb` file with `$fdisplay`.
- **Line 11**—Write to both files. This works because the call to `$fdisplay` uses the merged MCD.
- **Line 12**—Always close files when you are done with them.

All of this MCD work is meaningless unless we turn on the `OVM_LOG` action on one or more of our reporting functions. We'll see how to set actions below.

13.3 Controlling OVM Reporting

We control OVM reporting by calling methods in `ovm_top`. There are three types of settings:

- **Global settings**—These settings affect all the message calls.
- **Action settings**—Control what actions happen based on the severity, the ID, or a combination of both.
- **File settings**—Control where the logging output goes based on the severity, the ID, or a combination of both.

All of the methods we discuss in this section are methods in the `ovm_top` object. So, though we don't show it in the tables, always call these methods by referencing `ovm_top` like this:

```
ovm_top.report_header()
```

13.3.1 Global Reporting Methods

The global reporting methods control all four reporting functions in `ovm_top`. Consequently, they apply regardless of the ID or severity.

TABLE 13-2. Global Reporting Controls

Method	Behavior
`die()`	Finishes the current simulation at the end of the current simulation time slot and prints a summary.
`dump_report_state()`	Prints out all the current action, severity, file relationships.
`report_header()`	Prints a nifty header showing the OVM version.
`report_summarize()`	Shows the count of all messages by severity and ID.
`reset_report_handler()`	Sets `ovm_top` back to the default state.
`set_report_default_file (int mcd)`	Sets the file that receives all log entries that have not been specified to go elsewhere.
`set_report_max_quit_count (int max_count)`	Enables error counting and sets the maximum number of errors. The simulation calls `ovm_top.die()` when the count reaches this number.
`set_report_verbosity_level (int verbosity_level)`	Changes the verbosity level for the simulation.

13.3.2 Controlling Actions per Severity and ID

You can control what happens when you call a reporting method by creating an action, as we discussed in Section 13.2.1 on page 156, and then passing it to one of these methods.

The reporting methods choose their actions based on the message ID, the severity level, or both. The message IDs are strings. The severity levels must be one of four values: `OVM_INFO`, `OVM_WARNING`, `OVM_ERROR`, or `OVM_FATAL`.

TABLE 13-3. Summary of `ovm_top` Reporting Control Methods

Method	Behavior
`set_report_severity_action (severity s, action a)`	Changes the action based on the report severity.
`set_report_id_action (string id, action a)`	Sets the action based on the message ID.
`set_report_severity_id_action (severity s, string id, action a)`	Changes the action based on a combination of the report severity and the message ID.

13.3.3 Controlling Files per Severity and ID

If you used one of the action methods above to set an action to `OVM_LOG`, then you need to tell the OVM where to log the messages. You do that with one of the methods below.

All these methods take an MCD like the ones we discussed in Section 13.2.2 on page 157. They also take either a message ID, a severity level, or both. The message IDs are strings. The severity levels must be one of four values: `OVM_INFO`, `OVM_WARNING`, `OVM_ERROR`, or `OVM_FATAL`.

TABLE 13-4. Summary of `ovm_top` Reporting Control Methods

Method	Behavior
`set_report_default_file (integer mcd)`	Sets the file that receives all log entries that have not been specified to go elsewhere.
`set_report_severity_file (severity s, integer mcd)`	Changes the log file based on the report severity.
`set_report_id_file (string id, integer mcd)`	Sets the logging file based on the message ID.
`set_report_severity_id_file (severity s, string id, integer mcd)`	Changes the log file based on a combination of the report severity and the message ID.

13.3.4 Example of Reporting

Let's say that we want to log error messages into a file, but only if they are from ID `TOP`. We can make that happen with the following code:

```
1  import ovm_pkg::*;
2
3  module top;
4
5   initial begin
6     integer err_mcd;
7     err_mcd = $fopen("errors.txt");
8     ovm_top.set_report_verbosity_level(1000);
9     ovm_top.set_report_severity_id_action(OVM_ERROR, "TOP", OVM_LOG);
10    ovm_top.set_report_severity_id_file  (OVM_ERROR, "TOP", err_mcd);
11    ovm_top.dump_report_state();
12
13    ovm_top.ovm_report_info("TOP", "This won't go in a file", 300);
14    ovm_top.ovm_report_error("TOP", "This will only go in a file",300);
15    ovm_top.ovm_report_error("NOTTOP", "This won't go in a file either",300);
16
17    $fclose(err_mcd);
18  end
19 endmodule
```

FIGURE 13-5. Simple Reporting Example

- **Line 1**—We need the `ovm_pkg` to access `ovm_top`.
- **Line 6**—We store our MCD in the `err_mcd` integer.
- **Line 7**—Open the `errors.txt` file for writing by using `err_mcd`.
- **Line 8**—Set `ovm_top` to print all messages with a verbosity below 1000.
- **Line 9**—Tell `ovm_top` to log all messages written with an ID of "`TOP`" and a severity of `OVM_ERROR`.
- **Line 10**—Tell ovm_top to use `err_mcd` as the file handle for all messages with an ID of `TOP` and a severity of `OVM_ERROR`.
- **Line 11**—This shows us the state of `ovm_top`.
- **Line 13**—Print a message with an ID of `TOP` but a severity of `OVM_INFO`. This prints to the screen.
- **Line 14**—Print a message with an ID of `TOP` and a severity of `OVM_ERROR`. This goes into our file and doesn't print to the screen.
- **Line 15**—Print a message with an ID of `NOTTOP` and a severity of `OVM_ERROR`. This prints to the screen.
- **Line 16**—Clean up after ourselves by closing the error file.

When we run this short program, we get a large amount of output. Most of it comes from dumping the state of `ovm_top` on line 11 in Figure 13-5 on page 162.

When we open `errors.txt` we see what we expect:

```
OVM_ERROR @ 0 [TOP] This will only go in a file
```

FIGURE 13-6. Our `errors.txt` File Looks OK

The information we log to a file looks exactly like the information we would have displayed on the screen. It contains the severity, the ID, and then our message. Notice that because we didn't pass the `` `__FILE__ `` or `` `__LINE__ `` macros to this reporting method, we don't see that information in this file.

The `dump_report_state()` method on line 11 is our `ovm_top` debugging tool. It shows use the current state of all our message IDs, as well as the severities, the actions, and files associated with each of them.

When we call `dump_report_state()`, we get a large screen report that is broken into sections for actions, files, and counts.

This output gives us a complete picture of the reporting state inside `ovm_top`:

```
29  # ---------------------------------------------------------------
30  # OVM-2.0
31  # (C) 2007-2008 Mentor Graphics Corporation
32  # (C) 2007-2008 Cadence Design Systems, Inc.
33  # ---------------------------------------------------------------
34  # ---------------------------------------------------------------------
35  # report handler state dump
36  #
37  # max verbosity level =        1000
38  #
39  # +-------------+
40  # |   actions   |
41  # +-------------+
42  #
43  # *** actions by severity
44  # OVM_INFO = DISPLAY
45  # OVM_WARNING = DISPLAY
46  # OVM_ERROR = DISPLAY COUNT
47  # OVM_FATAL = DISPLAY EXIT
48  #
49  # *** actions by id
50  #
51  # *** actions by id and severity
52  # OVM_ERROR:TOP --> LOG
53  #
54  # +-------------+
55  # |    files    |
56  # +-------------+
57  #
58  # default file handle =           0
59  #
60  # *** files by severity
61  # OVM_INFO =           0
62  # OVM_WARNING =           0
63  # OVM_ERROR =           0
64  # OVM_FATAL =           0
65  #
66  # *** files by id
67  #
68  # *** files by id and severity
69  # OVM_ERROR:TOP -->           2
70  # report server state
71  #
72  # +-------------+
73  # |   counts    |
74  # +-------------+
75  #
76  # max quit count =      0
77  # quit count =      0
78  # OVM_INFO :     0
79  # OVM_WARNING :     0
80  # OVM_ERROR :     0
81  # OVM_FATAL :     0
82  # ---------------------------------------------------------------------
83  # OVM_INFO @ 0 [TOP] This won't go in a file
84  # OVM_ERROR @ 0 [NOTTOP] This won't go in a file either
```

FIGURE 13-7. The State of `ovm_top`

- **Line 37**—The maximum verbosity is, indeed, 1000.
- **Lines 43–47**—The actions associated with each severity level are still at the defaults.

- **Line 49**—There are no actions assigned to an ID by itself.
- **Line 52**—If we have a severity of `OVM_ERROR` and an ID of `TOP,` we've set the action to log data.
- **Line 58**—There is no default MCD for the file handle.
- **Lines 61–64**—There are no MCDs assigned to each severity level by itself.
- **Line 67**—There are no MCDs assigned to any IDs by themselves.
- **Line 69**—We've assigned an MCD to a severity of `OVM_ERROR` and an ID of `TOP`. Notice that this is a one-bit number. If we had multiple MCDs, we could OR them together to write to multiple files.
- **Line 76**—We did not set a maximum quit count, so no counting happens.
- **Lines 78–81**—We haven't written any messages, so all these numbers are zero.
- **Lines 83–84**—Here are our two messages that weren't logged to a file.

13.4 Summary

In this chapter we learned how to use `ovm_top` to control the output from our simulations.

We saw that we could control the number of messages with the verbosity functionality in the basic reporting methods. Then we saw how to define advanced actions in our reporting calls.

We learned about multichannel descriptors (MCDs) and how to use them when we open files for writing.

We learned how to control the reporting behavior of `ovm_top` with methods, and how to filter our actions and outputs based on the severity level of our messages and the ID in our message calls.

In our next chapter, we'll learn how to move transactions through the test bench.

14 Moving Transactions in Test Bench

IN THIS CHAPTER:

- *A Handle Hazard*
- *Safe Handle Handling*

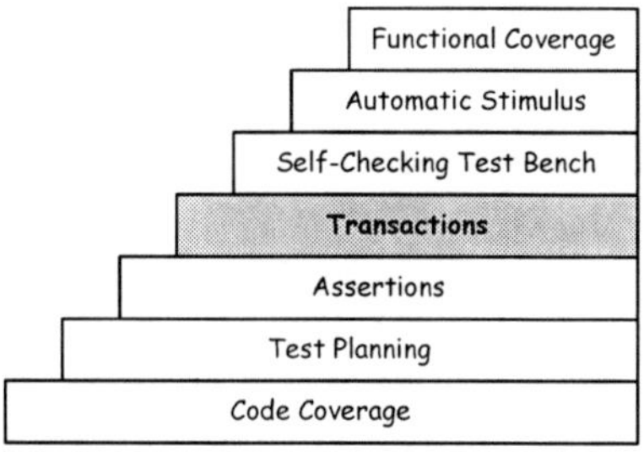

Transaction-level modeling (TLM) has existed in the industry for over fifteen years. TLM refers to a test bench where we encapsulate the complexity of signal-level DUT communication so that it becomes easy to write tests at a higher level.

However, TLM requires computer languages that can create objects. Verilog and VHDL can't do that, so until now all TLM test benches were written in C++, Vera, or *e*.

These languages can create objects, but they force the engineer to think in a different language. More importantly, they required that a hardware engineer think like a software engineer. Because it's difficult to switch disciplines, few hardware engineers would try to write TLM test benches.

That problem no longer exists, because now we have SystemVerilog—the TLM language for the rest of us. SystemVerilog allows us to allocate memory like C++ or *e*, but it doesn't force us to think like a software engineer. It doesn't make us master object-oriented programming or devise new concepts of what goes in the top level of a test bench.

SystemVerilog allows us to think like hardware engineers, creating test benches made of modules with connections between them. The TLM difference is that we connect the modules using `tlm_fifos` rather than wires. This approach to test bench creation is much easier for hardware engineers to implement than object-oriented software approaches.

All the components in a transaction-level test bench are either modules or components depending upon whether we are talking about SystemVerilog, Verilog, or VHDL. There are four types of blocks at the top level of an transaction-level test bench:

- **Transaction-level modules**—These modules have references only to `tlm_fifos` on their port list. They only read or write transactions to `tlm_fifos`.
- **Driver modules**—These modules have one or more `tlm_fifos` on their port list and RTL signals. They implement input channels. They read transactions from their `tlm_fifos` and translate them to signals.
- **Responder modules**—These modules have one or more `tlm_fifos` on their port list and RTL-level signals. They implement output channels. They read from the RTL signals and convert the signals to transactions.
- **RTL blocks**—These blocks only have RTL signals on their port list. They are usually the device under test and are written in synthesizable RTL.

In this chapter, we examine transaction-level modules and communication through a `tlm_fifo`. In the next chapter we are going to examine driver and responder modules to create a complete test bench.

14.1 A Handle Hazard

In Chapter 11, we wrote a module that created transactions and printed them to the screen. We're going to break that behavior up between two modules now to show how transaction-level modules communicate.

14.1.1 The Top Level

The top level of this example has two transaction-level modules connected by a FIFO. Each module has its own thread. There are two modules in the design:

- `producer`—The `producer` creates the `alu_operation` transactions we defined in Section 11.3.2 on page 133 and writes them to the `tlm_fifo` called `fifo`. After it writes the transaction to the `tlm_fifo`, it modifies it and writes it again. It produces three transactions.

- `consumer`—The `consumer` loops forever. It waits ten clock cycles and then gets an `alu_operation` from `fifo`. It prints the operation and goes back to the top of the loop.

```
1  import ovm_pkg::*;
2  import tinyalu_pkg::*;
3
4  module top;
5
6    tlm_fifo #(alu_operation) fifo;
7
8    producer tester   (fifo);
9    consumer printer  (fifo);
10
11   initial
12     begin
13       ovm_top.set_report_verbosity_level(10000);
14       fifo = new("fifo");
15       fork
16         tester.run();
17         printer.run();
18       join_none
19     end
20
21 endmodule
```

FIGURE 14-1. Top-Level Code for Producer/Consumer

- **Lines 1–2**—We import `ovm_pkg` and `tinyalu_pkg` to get all the declarations and definitions we use in the top level.
- **Line 6**—We have a `tlm_fifo` called `fifo` which transports `alu_operation` objects.
- **Line 8**—Instantiate the `producer` and call it `tester` and connect it to `fifo`.
- **Line 9**—Instantiate the `consumer` and call it `printer` and connect it to `fifo`.
- **Line 13**—Set the verbosity level for this simulation. We do this explicitly for two reasons. First, it gives us control of the verbosity rather than relying on defaults. Second, a future engineer will know where to look to change the verbosity.
- **Line 14**—Create the `fifo` object before we launch the module threads.
- **Lines 15–18**—Launch the `run()` task from each module in its own thread.

This is all we want from the top level. Now it's up to the producer and consumer to do the work.

Notice that this test bench looks almost exactly like a test bench written entirely in VHDL or Verilog. There are modules and they communicate. There are only two significant differences:

1. The modules communicate by using a `tlm_fifo` rather than a wire.

2. We need to launch the multiple threads from our top level. In an RTL simulation, our `always` block or `process` block would launch itself.

We are passing each module a pointer to a `tlm_fifo` object that we create on the top level. The `producer` and `consumer` module both expect the variable `fifo` to contain a pointer to a `tlm_fifo`. Therefore, we need to make sure that we create that `tlm_fifo` before we launch the threads.

Remember, when you declare a variable that holds an object like a `tlm_fifo` or a transaction, you are not allocating the memory for that object. You are simply supplying a space where SystemVerilog will put the pointer to the object once it's created.

14.1.2 The Producer Module

The `producer` module implements a small test by doing the following:

- Creating a new transaction.
- Pass it to the `tlm_fifo`.
- Modify the transaction to create new stimulus.
- Pass the modified transaction to the `tlm_fifo`.

It passes three transactions to the `tlm_fifo`.

Here is the `producer.` This is a pure transaction-level module:

```
1  import ovm_pkg::*;
2  import tinyalu_pkg::*;
3
4  module producer (ref tlm_fifo #(alu_operation) fifo);
5
6    alu_operation t;
7    string m;
8
9    task run;
10
11     $sformat(m,"%m");
12
13     // First transaction
14     t = new (8'h00, 8'h00, add_op);
15     ovm_top.ovm_report_info(m, {"WILL BE LOST: ",t.convert2string});
16     fifo.put(t);
17
18     // Second transaction
19     t.A = 8'hFF;
20     t.B = 8'h55;
21     ovm_top.ovm_report_info(m, {"generated: ",t.convert2string});
22     fifo.put(t);
23
24     // Third transaction
25     t.A = 8'hAA;
26     t.B = 8'hEE;
27     t.op = xor_op;
28     ovm_top.ovm_report_info(m, {"generated: ",t.convert2string});
29     fifo.put(t);
30   endtask
31
32 endmodule
```

FIGURE 14-2. The TLM Producer Module

- **Line 4**—There is only a reference to a `tlm_fifo` on the module port list. This makes it a transaction-level module. The variable that holds our `tlm_fifo` is called `fifo` (not very original, I know.)
- **Line 6**—Declare the transaction that we'll pass into the test bench.
- **Line 7**—The string `m` holds the path to this module. We'll use it as an ID in our calls to the OVM reporting methods.
- **Line 9**—We do everything in a `run()` task because this is a transaction-level module.
- **Line 11**—Before we start, store the path to the module in `m`.
- **Line 14**—Create a new transaction
- **Line 15**—Print it out.
- **Line 16**—Put it into `fifo`.
- **Lines 19–20**—Modify the transaction
- **Line 21**—Print it out. Notice the use of the `convert2string()` method we wrote when we defined the transaction.
- **Line 22**—Put it into `fifo`.

- **Lines 25–29**—Put a third transaction into `fifo`.

14.1.3 The Consumer Module

The `consumer` module simply reads a transaction from `fifo` and prints it to the screen using `ovm_top.ovm_report_info()` and the `convert2string()` method. The top of this module is the same as the consumer. The only difference is in the `run()` task:

```
1  import ovm_pkg::*;
2  import tinyalu_pkg::*;
3
4  module consumer(ref tlm_fifo #(alu_operation) fifo);
5
6    alu_operation op;
7    string m;
8    task run;
9      $sformat(m, "%m");
10     forever begin
11       #10;
12       fifo.get(op);
13       ovm_top.ovm_report_info(m,{"Received: ", op.convert2string,"\n"});
14     end
15   endtask
16 endmodule
```

FIGURE 14-3. TLM Consumer Module

- **Line 10**—This task loops forever.
- **Line 11**—Wait 10 time units to simulate RTL running.
- **Line 12**—Get the transaction from the `tlm_fifo` called `fifo`.
- **Line 13**—Print it out.

That's all there is to the `consumer`.

14.1.4 Running Our Example

When we run our simulation we get the following (erroneous) results:

```
40  # ---------------------------------------------------------------
41  # OVM-2.0
42  # (C) 2007-2008 Mentor Graphics Corporation
43  # (C) 2007-2008 Cadence Design Systems, Inc.
44  # ---------------------------------------------------------------
45  # OVM_INFO @ 0 [top.tester.run] WILL BE LOST: A: 00  B:00   OP:add_op
46  # OVM_INFO @ 0 [top.tester.run] generated: A: ff  B:55   OP:add_op
47  # OVM_INFO @ 10 [top.printer.run] Received: A: ff  B:55   OP:add_op
48  #
49  # OVM_INFO @ 10 [top.tester.run] generated: A: aa  B:ee   OP:xor_op
50  # OVM_INFO @ 20 [top.printer.run] Received: A: aa  B:ee   OP:xor_op
51  #
52  # OVM_INFO @ 30 [top.printer.run] Received: A: aa  B:ee   OP:xor_op
```

FIGURE 14-4. An Error in Our Simple Example

This output shows us that the `producer` (instance name `top.tester`) created three transactions and the `consumer` (instance name `top.printer`) read three transactions.

Unfortunately, there is a problem. While the `producer` created three different transactions, the `consumer` printed only two different transactions. The transaction generated at time 10 (line 49) gets printed twice, and the transaction generated at time 0 (line 45) didn't get printed at all.

What happened?

14.2 Safe Handle Handling

In the previous section we created a simple `producer/consumer` pair of modules and immediately ran into trouble. Our attempt to get the `producer` to feed three different transactions to our `consumer` seems to have failed, because the `consumer` is printing out only two different transactions.

This problem demonstrates a common pitfall in transaction level modeling. What caused us to lose a transaction in our previous example? The answer lies in a clear understanding of how object variables work.

When you look at a declaration like this,

```
alu_operation op;
```

you may think that SystemVerilog is putting some space aside to store an `alu_operation` object in a variable called `op`. This is a common assumption among engineers accustomed to Verilog and VHDL. RTL languages always set aside space for a variable as soon as it is declared.

This is the wrong assumption, because SystemVerilog does not set aside the memory for an object until you construct it. The object variables are just handles to memory. In the above example, there is no memory set aside for `op` until we execute the expression `op=new("op")`.

This difference creates a problem for people who assume that each object variable has its own memory set aside. Consider this snippet:

```
alu_operation op1, op2;

op1 = new(8'hFF,8'h55,add_op);
op2 = op1;
```

We declared two `alu_operation` variables, and then we called `new()` and stored the value in one of them. Then we copied `op1` to `op2`.

We might think that this snippet gives us two copies of the `alu_operation` object that was put into `op1`— but it doesn't. Instead, the assignment does exactly what it says it does: it copies the pointer from one variable to another.

It's easier to imagine the situation with a picture. If we draw the variables as small boxes, and then make a larger box that holds the memory with the transaction, the situation becomes clear.

Here is what we are doing, in pictures:

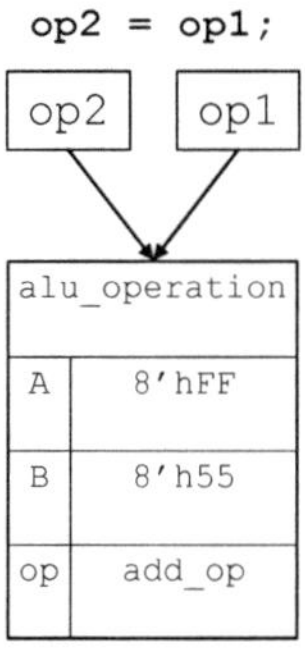

FIGURE 14-5. Result of Copying Handles

Now we have two variables pointing to the same object. This means that if we change `op1`, we also change `op2`. Surprise!

If we want to avoid this problem, we must explicitly make a clone of the object when we copy it. This is why we created that `clone()` method back when we defined the `alu_operation` object (See Figure 11-5, "Methods for `alu_operation`," on page 134).

The `clone()` method creates a new object, and thus allocates new memory. Then it copies the values from the first object into the new memory for the second object. When you clone an object, you are creating an entirely new object to store in the new location.

Here's what happens when we use `clone()`:

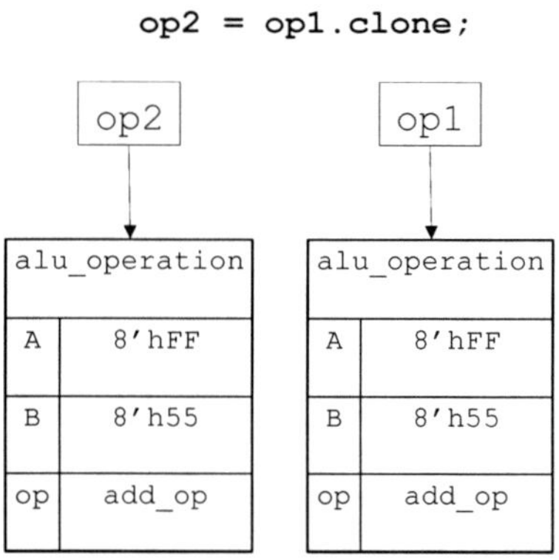

FIGURE 14-6. Results of Cloning a Transaction

Now we have two discrete copies of the object. Changes we make to the object pointed to by `op2` do not affect `op1`.

When we move transactions around a test bench, we need to clone them to avoid the problems we saw with our `producer/consumer` example. Which brings us to the question, what happened to our `producer/consumer` example?

14.2.1 Fixing the Bug in `producer/consumer`

The `producer` caused the bug that we saw because it didn't clone the transaction when it put it into the `tlm_fifo`. If we look at the code in Figure 14-2 on page 171, we see that we created a new object in line 14 and put a copy of the handle to that object into `fifo` on line 16.

Then we modified the transaction using the variable `t`. The problem is that `t` was still pointing to the object that we had just put into `fifo`. So when we changed `t`, we were really changing the transaction in `fifo`. The data from the first transaction was lost.

There are two ways to fix the bug:

1. We could have called `new()` each time wanted a new transaction. That way we'd have a new pointer and the transaction in `fifo` would be safe.
2. We could have cloned the transaction when we put it into `fifo` using the `clone()` method. This would have created a new transaction to store in `fifo`.

Let's fix the producer by using the `clone()` method:

```
1  import ovm_pkg::*;
2  import tinyalu_pkg::*;
3
4  module producer (ref tlm_fifo #(alu_operation) fifo);
5
6    alu_operation t;
7    string s;
8
9    task run;
10
11     $sformat(s,"%m");
12     t = new (8'h00, 8'h00, add_op);
13     ovm_top.ovm_report_info(s, {"generated: ",t.convert2string});
14     fifo.put(t.clone);
15
16     t.A = 8'hFF;
17     t.B = 8'h55;
18     ovm_top.ovm_report_info(s, {"generated: ",t.convert2string});
19     fifo.put(t.clone);
20
21     t.A = 8'hAA;
22     t.B = 8'hEE;
23     t.op = xor_op;
24     ovm_top.ovm_report_info(s, {"generated: ",t.convert2string});
25     fifo.put(t.clone);
26   endtask
27 endmodule
```

FIGURE 14-7. The Fixed `producer` with `clone()`

The new `producer` clones the transaction when it puts it into the `tlm_fifo` on lines 14, 19, and 25.

Now the simulation works properly because we've fixed our TLM communication problem:

```
# -----------------------------------------------------------------
# OVM-2.0
# (C) 2007-2008 Mentor Graphics Corporation
# (C) 2007-2008 Cadence Design Systems, Inc.
# -----------------------------------------------------------------
# OVM_INFO @ 0 [top.tester.run] generated: A: 00  B:00   OP:add_op
# OVM_INFO @ 0 [top.tester.run] generated: A: ff  B:55   OP:add_op
# OVM_INFO @ 10 [top.printer.run] Received: A: 00  B:00   OP:add_op
#
# OVM_INFO @ 10 [top.tester.run] generated: A: aa  B:ee   OP:xor_op
# OVM_INFO @ 20 [top.printer.run] Received: A: ff  B:55   OP:add_op
#
# OVM_INFO @ 30 [top.printer.run] Received: A: aa  B:ee   OP:xor_op
```

FIGURE 14-8. Properly Running TLM Communication Code

14.2.2 Choosing When to Clone

Always cloning our transactions is the safest way to use a `tlm_fifo`, but it can also be a waste of memory. If our transaction contains something large, such as a video frame, we may not want to make new copies of it whenever we move it between modules.

In general, we don't need to clone a transaction if we can be sure that we'll never modify it again. For example, if we had created three new transactions in our `producer` rather than modifying the one we had, we never would have seen a problem in our design. The consumer would have received three different handles.

We always want to clone our transactions if we are putting them into two different `tlm_fifos`. We'll see this situation when we start creating self-checking test benches. This is because we can't be sure of what the downstream modules do. If we give two modules pointers to the same transaction, and one of the modules modifies the transaction, then the other module will see corrupted data. That can't happen if we clone the transactions we send to both modules.

In general, if your transactions are small and they won't eat up all your computer's memory, you are always safer cloning transactions as you put them into `tlm_fifos`. "Safer" means you are unlikely to spend time debugging strange situations where your transactions seem to be changing randomly.

14.3 Summary

In this chapter we learned that SystemVerilog gives us transaction-level communication without forcing us to think like a software designer. It does this with modules that talk to each other by using `tlm_fifo` objects.

We discussed objects and handles, and we saw that copying an object's handle is not the same thing as creating a new object with the same data. When we copy object handles, we get two pointers to the same object. We saw that we can get bugs in our test bench if we don't understand this behavior.

We avoided multiple pointers to the same object by using the `clone()` method. We saw that while we don't always need to clone our transactions, it is the easiest way to avoid confusing bugs. If we have transactions that take up a lot of memory, then we could easily create simulations that swamp our computer if we clone everything—but if we have small transactions, cloning is best.

Now we are ready to add drivers, responders, and an RTL DUT to our test bench. We'll do that in the next chapter.

15 Creating an RTL Test Bench

IN THIS CHAPTER:

- *The RTL Test Bench*
- *Writing a Driver*
- *Writing a Responder*
- *Checking Results*

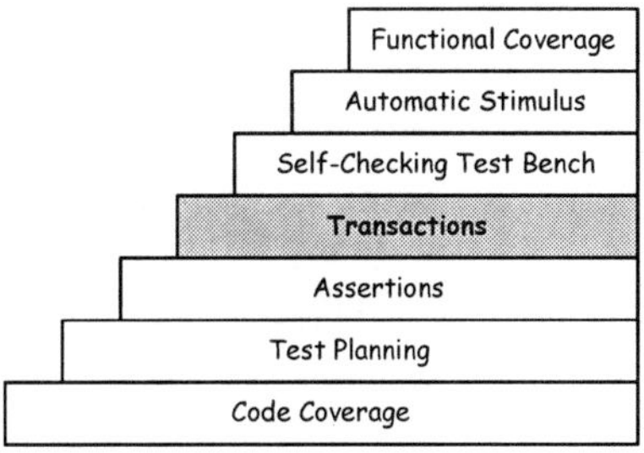

We're ready to build a complete transaction-level test bench for the TinyALU. Transaction-level test benches make it easy to create tests and review the results, because all the details of communicating with the DUT are hidden in modules.

This is the culmination of the test plan approach we used back in Chapter 3. We wrote that test plan in terms of input and output channels on our DUT, and then we described transactions within those channels. We captured the DUT's functionality in terms of transactions.

Now we complete the job by implementing the channels with `tlm_fifos` and driver and responder modules. The input channels use a `tlm_fifo` with a driver, and the output channels use a `tlm_fifo` with a responder.

The TinyALU has only one input channel and one output channel. (There are larger examples on www.fpgasimulation.com.) We'll now complete its test bench by creating the driver and responder for its input and output channels.

15.1 The RTL Test Bench Architecture

In the previous chapter, we used `tlm_fifos` to communicate between modules. This was the essence of the transaction-level test bench. We worked on a tester and a printer in that chapter. We're now going to expand that test bench to include a driver and a responder, along with a protocol monitor.

Here is the top level of the TinyALU's test bench:

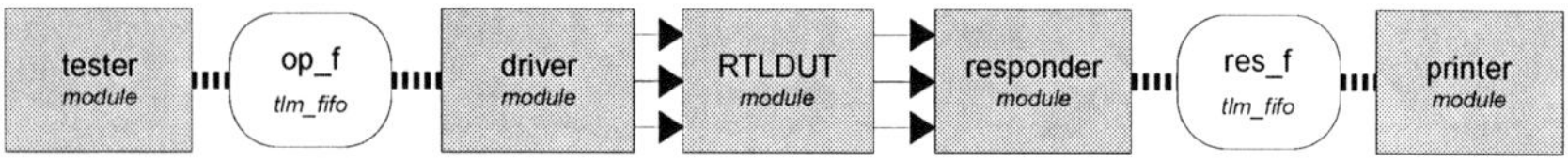

FIGURE 15-1. Transaction-Driven RTL Test Bench

There are five modules and two `tlm_fifos` in this design. The `op_f` FIFO moves transactions from the `tester` to the `driver`. The `res_f` FIFO moves transactions from the `responder` to the `printer`. Here are the five modules, identified by their instance names:

- `tester`—Creates tests by creating a stream of transactions.
- `driver`—Converts the `alu_operation` transactions to signals on the input pins of the TinyALU. It prints the transactions to the screen and into a file called `results.txt`.
- `RTLDUT`—This is the VHDL RTL version of the TinyALU. This is what we are testing.
- `responder`—Converts signals from the TinyALU to `alu_result` transactions.
- `printer`—Prints the result transactions to the screen and to a file called `results.txt`.

Notice that there are no arrows in the connections between the `tlm_fifos` and the modules. This is because the data in the connections can go either way. The driver could write to `op_f`, but we never ask it to.

We are going to examine this test bench one module at a time, and then put all the modules together in the end by means of the top level.

15.2 The tester Module

The `tester` module shows how easy it is to write tests when you are working at the TLM level. In this module we explicitly create four transactions with data that we define, and then we create 11 additional transactions with random data to reach a total of 15.

Here is the `tester`:

```
1  import ovm_pkg::*;
2  import tinyalu_pkg::*;
3
4  module generator(ref tlm_fifo #(alu_operation) op_f);
5
6  alu_operation op;
7
8     task run;
9         op = new (8'hFF, 8'h01, add_op);
10        op_f.put(op);
11        op = new (8'hFE, 8'h03, mul_op);
12        op_f.put(op);
13        op = new (8'h55, 8'hFF, and_op);
14        op_f.put(op);
15        op = new (8'h55, 8'hFF, xor_op);
16        op_f.put(op);
17        op = new();
18        for (int i = 1; i<=10; i++) begin
19           assert(op.randomize());
20           op_f.put(op.clone());
21        end
22        #500;
23        ovm_top.die();
24   endtask // run
25 endmodule
```

FIGURE 15-2. TinyALU Transaction-Level Test Generator

- **Line 4**—The generator connects to a `tlm_fifo` that takes `alu_operation` transactions.
- **Line 6**—This handle holds our transactions before we send them into the test bench.
- **Line 8**—The `run()` task. The generator is a transaction-level module, so it has a `run()` task.
- **Line 9**—Create a new transaction with an `add_op` operation.
- **Line 10**—Pass the new transaction to the test bench. Notice that we don't clone the transaction here. That's because we won't be touching this transaction again.
- **Lines 11–16**—We create new transactions for each of the other operations. Each call to `new()` gives us another block of memory that contains another object. Consequently, we don't need to clone these handles as we pass them into the `tlm_fifo`.

- **Lines 17–21**—Here we create one new transaction and then randomize it 11 times. Randomizing the transaction does not give us a new block of memory, but rather just changes the values in our object. As a result, we need to clone our object as we put it into `op_f`. Otherwise, the downstream modules would have a pointer to a continuously changing object.
- **Line 22–23**—Wait 500 units after the last transaction goes into the test bench, and then shut down the simulation. The `ovm_top.die()` method closes all the files and shuts down the simulation at the end of a simulation cycle.

The generator demonstrated the power of transaction-level test benches. It's easy to create new transactions without having to worry about how they'll get translated into signals.

Now we translate our input and output channels into modules. Let's start with the output channel. It's simpler.

15.3 The `responder` Module

In general, responder modules connect the signal world of the RTL DUT to the transaction world of the test bench. They understand the output protocol from the DUT, and use the information from the signals to create transactions on the output channels.[1]

Responder modules don't have a `run()` task. Instead, they use the `always` block construct to create a process that is sensitive to the clock in the design. In the case of the TinyALU, we use the positive edge of the clock for our responder module, because the TinyALU delivers results on a positive clock edge.

The TinyALU has one output transaction: `alu_result`. The `alu_result` transaction contains a 16-bit number. We implemented the `alu_result` transaction with an SystemVerilog class called `alu_result`. We defined the `alu_result` object in `tinyalu_pkg.sv`. Its code is in Figure 11-6 on page 135.

When we wrote our TinyALU test plan, we defined the input and output channels down to the waveform level. Initially, we used those waveforms to write tasks or

1. When we get to Step 5, Self-Checking Test Benches, we'll see that the transactions from the responder modules are the key to creating a self-checking test bench.

procedures that implemented the transaction at the signal level. Now we'll use the waveform to implement the TinyALU responder.

Here is the waveform from the TinyALU output channel:

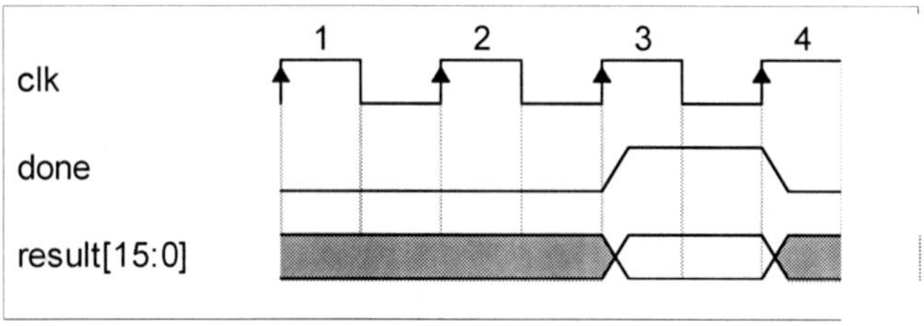

FIGURE 15-3. TinyALU Output Channel Waveform

If the `done` signal is high, then the `result` bus is valid. That's all there is to this protocol.

Our `alu_responder` module uses the positive edge of the clock for synchronization. Then it checks the `done` signal. If the `done` signal is high (and we are not in reset), the module takes the data from the result bus and uses it to create a new `alu_result` object.

Here is the `alu_responder` code:

```
16 import ovm_pkg::*;
17 import tinyalu_pkg::*;
18
19
20 module alu_responder (ref tlm_fifo #(alu_result) res_f,
21    input bit clk,
22    input bit reset_n,
23    input bit done,
24    input logic [15:0] result);
25
26    alu_result rslt;
27    string m;
28
29    initial $sformat (m,"%m");
30
31    always @(posedge clk)
32       if (done && reset_n) begin
33         rslt = new (result);
34         assert (res_f.try_put(rslt)) else
35           ovm_top.ovm_report_fatal(m,"res_f is full",0,`__FILE__,`__LINE__);
36         end
37    endmodule
```

FIGURE 15-4. TinyALU Responder Module

- **Line 20**—Like all converter modules, `alu_responder` has a reference to a `tlm_fifo` on its module port, along with signals from the RTL DUT. The `tlm_fifo` works with `alu_result` transactions. This class was defined in Figure 11-6 on page 135
- **Lines 21–24**—Here are the RTL-level signals on the TinyALU output.
- **Line 26**—Declare a variable to hold our `alu_result` transaction.
- **Line 31**—This module works on the positive edge of the clock.
- **Line 32**—Check to see whether the TinyALU has raised the `done` signal and that we are not in reset.
- **Line 33**—We use the `result` bus as an argument when we call the `new()` method. This creates a new `alu_result` transaction that contains the number currently on the `result` bus.
- **Line 34**—We try putting the transaction into `res_f`. We know that this `tlm_fifo` should never be full, because the `result_printer` should read it long before the next positive edge of the clock. Therefore, this `try_put()` can fail only if there is a fatal problem in the test bench.

The `responder` is converting between the clock-synchronized world of RTL and the `tlm_fifo`-synchronized world of TLM. This conversion means that we can't use the familiar `put()` and `get()` methods to interact with the `tlm_fifo`. We need to use `try_put()` and `try_get()`.

15.3.1 The try_put() and try_get() Methods

When we work at the transaction level, we create run() tasks and launch them in threads by using the fork...join_none construct. We synchronize between the threads by using the put() and get() methods in the tlm_fifo. These methods synchronize the threads by suspending on a get() if there is no data, and by suspending on a put() if there is no room in the tlm_fifo.

Though we still need to interact with the tlm_fifo from our always block, we can't use put() and get() because these methods would destroy our synchronization. This is because we will miss clock edges if we suspend on these methods. Instead of being synchronized to the clock edge, we'd be synchronized to some weird combination of clock edges and tlm_fifo suspensions. That would be random and bad.

We solve this problem by using the try_put() and try_get() methods when we are in a clock-synchronized always block. These methods never suspend. Instead, they return a 1'b1 if they were successful or a 1'b0 if they were unsuccessful. This allows us to test them and then go back to wait for the edge of the clock again.

We did something a little different with try_put() in our responder. There should always be room in res_f, so we know that a failed put is a fatal problem. We use an assert statement to test try_put(). If the res_f is full, we call ovm_report_fatal() and stop the simulation.

When we look at the driver module, we'll use try_get() instead of try_put()—but the is the same.

15.4 The driver Module

The TinyALU input channel is a little more complicated than the output channel. There are operations that can take multiple cycles to complete. Therefore, we need to implement a handshake where we raise the start signal to start an operation and wait for the done signal to see that the operation is completed.

The TinyALU has one transaction with several fields. Each transaction has the following data:

- **A and B operands**—These are the two eight-bit operands.
- **Operation**—The operation can be ADD, AND, XOR, MUL, or NOP.

This abstract description of the input transaction needs to be made more concrete for our test bench. We translate the transaction itself into a SystemVerilog class called `alu_operation`. The definition is in `tinyalu_pkg`. We discussed this definition in Figure 11-4 on page 133

We captured the channel waveform as part of our test plan:

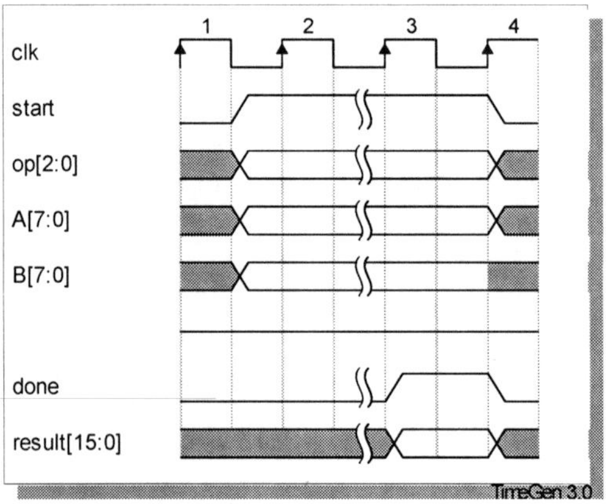

FIGURE 15-5. TinyALU Waveform

The `driver` outputs the signals `clk`, `start`, `op`, `A`, and `B`. The protocol says to drive these signals with an operation and keep them driven until the `done` signal goes high.

The `driver` also implements a delay between transactions. In the real world, a TinyALU would not get bombarded with operation after operation. There would be delays between the time that the `done` signal goes high and the `start` signal is raised again for the next operation. The `driver` inserts delays between operations to represent this.

Drivers are usually state machines, and complicated protocols have large complicated state machines. In our case, we have a very simple state machine. We are either in the middle of an operation or waiting for an operation. The `start` signal tells us which. If the `start` signal is high, then we are in the middle of an operation and we should look for the `done` signal. If it's not high, then we should find a new transaction and drive it into the TinyALU.

This gives us the following flow chart for the driver. This flow chart runs on the negative edge of every clock:

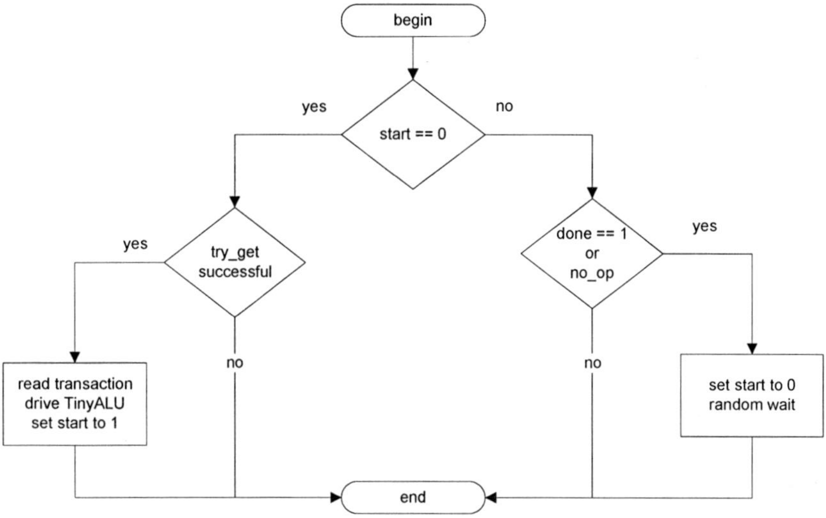

FIGURE 15-6. Flow Chart for TinyALU Driver Code

The `driver` checks to see if we are currently doing an operation. If the `start` signal is zero, then we are not doing an operation and we try to get a new transaction. If we successfully get a new transaction, we drive its data into the TinyALU.

If we are currently executing an operation, then we must check to see whether the TinyALU has raised the `done` signal. If we are done (or if this operation was a `no_op`), then we set the `start` signal to zero. Finally, we wait some random number of clocks and look for the next transaction.

There are two parts to the driver's code. The first part, basic housekeeping, implements the signal interface and also creates the clock and initial reset. The second part implements the state machine.

Here's the first part:

```
16  import ovm_pkg::*;
17  import tinyalu_pkg::*;
18  `include  "tinyalu_defines.svh"
19
20  module alu_driver (
21      ref tlm_fifo #(alu_operation) op_f,
22      output bit clk,
23      output bit reset_n,
24      input  bit done,
25      output bit start,
26      output logic [2:0] op,
27      output logic [7:0] A,
28      output logic [7:0] B);
29
30     alu_operation request;
31     string m = "";
32
33     initial $sformat(m,"%m");
34
35     initial begin : clockgen
36        clk = 0;
37        forever #10 clk = ~clk;
38     end
39
40     initial begin : resetgen
41        reset_n = 0;
42        @(posedge clk);
43        @(negedge clk);
44        reset_n = 1;
45     end
46
```

FIGURE 15-7. Driver Housekeeping

- **Line 18**—This header file contains the macros that control writing output to a data file.
- **Lines 20–28**—The driver connects to a `tlm_fifo` called `op_f` that transfers `alu_operations`. It also connects to all the signals on the TinyALU DUT.
- **Lines 35–37**—The driver provides the clock. Finally, after all this high-level transaction work, we have a real clock!
- **Lines 40–45**—This block resets the DUT. It lowers the `reset_n` signal and waits a clock cycle and a half to raise it again.

Now that the driver module is set up with a clock and a reset, we can start running the algorithm in our flow chart. The flow chart runs in an `always` block that waits for the negative edge of the clock.

Here's the state-machine code:

```
46
47     always @(negedge clk) begin : drive_transactions
48       if (!reset_n) begin : handle_reset
49         start = 1'b0;
50       end else
51         if (!start) begin : start_operation
52            if (op_f.try_get(request)) begin : new_transaction
53              ovm_top.ovm_report_info(m,          request.convert2string(),300);
54              ovm_top.ovm_report_info(`OUTPUTID,request.convert2string(),  0);
55              op = request.op2bits();
56              A = request.A;
57              B = request.B;
58              start = 1'b1;
59            end
60          end
61        else begin : in_operation
62          if (done || (op == 3'b000)) begin : end_operation
63            @(posedge clk);
64            start = 0;
65            wait_rand_clocks;
66          end
67        end
68      end
69
70     task wait_rand_clocks;
71        logic [1:0] waittime;
72        waittime = $random;
73        repeat (waittime) @(posedge clk);
74     endtask
75  endmodule
```

FIGURE 15-8. Driver State Machine

- **Line 47**—This `always` block creates our thread the old-fashioned way. It runs on the negative edge of the clock.
- **Line 51**—Check to see if we are in the middle of an operation. If we aren't, then we're going to get a transaction and drive the bus.
- **Line 52**—Try to get a transaction from `op_f`.
- **Line 53**—Print the transaction to the screen with a verbosity of 300. We can turn off these messages if we want the simulation to run faster.
- **Line 54**—Print the transaction into the output file. The macro `'OUTPUTID` contains the message string that is connected to the results file. We are assuming that the top-level module opened the file and connected it to `'OUTPUTID`.

  We set the verbosity to 0. This way we write to the file regardless of whether we write to the screen.
- **Line 55**—Drive the operation code onto the `op` bus. The `alu_operation` object stores the operation as an enumerated type. We can't drive an enumerated type onto a bus, so we created the `op2bits()` method to convert the enumerated type to the correct bits for the `op` bus.
- **Lines 56–57**—Drive the `A` bus and `B` bus with the operands. The `alu_operation` contains two eight-bit operands.

Line 58—Raise the `start` signal to start the TinyALU.

Line 61—If we are here, then the TinyALU is currently processing an operation.

Line 62—If the `done` signal has been raised, or the operation is a `no_op`, then we drop the start signal back to zero.

Line 65—The TinyALU should not be pummeled with back-to-back operations. We need to make sure it can wait between operations, so we have a task that waits a random number of clock cycles.[2]

Lines 70–74—This task fills a two-bit number with a random number from 0 to 3. Then it waits that many clock cycles before returning.

We've now got a test bench that reads easy-to-use transactions and translates them into signals. We also translate signals back into transactions. The only thing left is to write the transactions into a file.

15.5 Transaction Printer

The test bench we described in Figure 15-1 on page 182 created transactions, ran through the DUT, got transactions from the DUT, and then printed out the transactions.

The `printer` is a transaction-level test bench. It has no clock and synchronizes its `run()` method with a `tlm_fifo`. It reads the transactions from the `responder` and prints them.

The goal is to have one file that contains all the transactions that went into the test bench, as well as all the transactions that came out. Then we can use this file to determine whether our DUT is working properly.

2. We study `$random` in Chapter 19.

The printer uses the same macros as the top level and the driver module, so it writes the transactions to the same file as the driver. Here's the code:

```
16  import ovm_pkg::*;
17  import tinyalu_pkg::*;
18  `include  "tinyalu_defines.svh"
19
20  module result_printer(ref tlm_fifo #(alu_result) res_f);
21
22     alu_result restran;
23     string     filename;
24     string     m;
25     integer    file_handle;
26
27     task run;
28        $sformat(m,"%m");
29        forever begin
30           res_f.get(restran);
31           ovm_top.ovm_report_info(m,
32                 restran.convert2string(),300);
33           ovm_top.ovm_report_info(`OUTPUTID,
34                 restran.convert2string(),  0);
35         end
36     endtask
37  endmodule
```

FIGURE 15-9. Transaction-Level Printer Module for the TinyALU

- **Line 18**—This include file contains the definition of the `'OUTPUTID` macro. That macro contains the message ID that prints to a file.
- **Line 27**—This transaction-level module has a `run()` task.
- **Line 30**—Get a result transaction out of `res_f`.
- **Line 31**—Write the transaction to the screen with a verbosity of 300. We can turn off the messages to speed up simulation.
- **Line 32**—Write the transaction into the output file with a verbosity of 0. This way the file is written regardless of the verbosity level. The `'OUTPUTID` macro was defined in `tinyalu_defines.svh`.

Now all we need to do is put all the modules together at the top level.

15.6 The TinyALU Top Level

There are two parts to the top level of this test bench. In the first part, we instantiate all the modules that make up the test bench. In the second part, we set up the output file, set the verbosity level, create the `tlm_fifos`, and launch the threads in the transaction-level modules.

Here's the first part of the top level, the module instantiations:

```
import    ovm_pkg::*;
import    tinyalu_pkg::*;
`include  "tinyalu_defines.svh"

module top;
   tlm_fifo #(alu_operation) op_f;
   tlm_fifo #(alu_result)    res_f;
   wire clk;
   wire reset_n;
   wire done;
   wire start;
   wire [2:0] op;
   wire [7:0] A;
   wire [7:0] B;
   wire [15:0] result;

   generator       tester      (.op_f(op_f));

   alu_driver driver
     (.op_f(op_f),
      .clk(clk),
      .reset_n(reset_n),
      .start(start),
      .done (done),
      .op(op),
      .A(A),
      .B(B));

   tinyalu RTLDUT
     (.clk(clk),
      .reset_n(reset_n),
      .start(start),
      .op(op),
      .A(A),
      .B(B),
      .done (done),
      .result(result));

   alu_responder responder
     (.res_f(res_f),
      .clk(clk),
      .reset_n(reset_n),
      .done(done),
      .result(result));

   result_printer printer   (.res_f(res_f));
```

FIGURE 15-10. Test Bench Module Instantiations

- **Line 32**—The `generator` is a transaction-level module, so it has only a `tlm_fifo` on its port list.
- **Line 34**—The `alu_driver` is a driver module, so it contains a `tlm_fifo` and the signals necessary to drive the DUT. This module creates the clock.
- **Line 44**—The `tinyalu` is an RTL-level design. It has only signals on its ports.
- **Line 54**—The `alu_responder` is a responder module, so it contains signals from the DUT and a `tlm_fifo` on its port list.
- **Line 61**—The `result_printer` is a transaction-level module, so it only has a `tlm_fifo` on its port list.

Once we've instantiated the modules, we are ready to set up the test bench and run it. Here is the code. This code runs only once to set up the simulation:

```
63   initial begin
64        integer output_mcd;
65        string m;
66        $sformat(m,"%m");
67        assert((output_mcd = $fopen(`OUTPUTFILENAME))) else
68          ovm_top.ovm_report_fatal(m,
69                   {"Could not open ", `OUTPUTFILENAME},
70                   0,`__FILE__, `__LINE__);
71        ovm_top.set_report_verbosity_level(0);
72        ovm_top.set_report_id_file  (`OUTPUTID, output_mcd);
73        ovm_top.set_report_id_action(`OUTPUTID, OVM_LOG);
74        op_f  = new("op_f");
75        res_f = new("req_f");
76        fork
77           tester.run();
78           printer.run();
79        join_none
80     end
81  endmodule
```

FIGURE 15-11. Test Bench Initial Block

- **Line 67**—We open the file name that was supplied in the `tinyalu_defines.svh` file. Alternatively, one could supply the file name on the simulation command line.[3]
- **Lines 68–70**—If we were unable to open the file, we report a fatal error.
- **Line 71**—Set the verbosity level to 0 so that we don't see messages printed to the screen. This speeds up the simulation.
- **Line 72**—Set the `ovm_top` object so that it prints all messages with the `'OUTPUTID` message ID to the file opened on line 67. This applies to all modules, so we can write to the same file from several places.

3. To learn how to do this, visit the forums on www.fpgasimulation.com.

- **Line 73**—Set the `ovm_top` object so that all messages with the `'OUTPUTID` message ID log their output to a file.
- **Lines 74–75**—Create the `tlm_fifo` objects before we start the threads.
- **Lines 78–79**—Launch the `run()` tasks in the `tester` and `printer` as threads to start the simulation.

The `driver`, `responder`, and `RTLDUT` start simulating immediately. Their `always` blocks or `process` blocks suspend as soon as they have to wait for the clock.

When we launch the `tester.run()` thread, it puts a transaction into `op_f` and the `driver` reads the transaction and starts the simulation. This kicks things off. The `tester` keeps loading transactions into `op_f` until it's done. Then it waits 500 ns and shuts down the simulation.

15.7 Running the TinyALU RTL Test Bench

Our test bench creates a file[4] that saves transaction requests and results. We used the OVM reporting tools to open this file, and then we wrote to it in the driver and printer modules.

The driver module wrote transaction requests (of type `alu_operation`) into the file, and the printer module wrote transaction results (of type `alu_result`.)

4. Called `results.txt` in the example that you can get from www.fpgasimulation.com.

The result looks like this:

```
1 OVM_INFO @ 20 [OUTPUT] A: ff  B:01   OP:add_op
2 OVM_INFO @ 50 [OUTPUT] Result: 0100
3 OVM_INFO @ 60 [OUTPUT] A: fe  B:03   OP:mul_op
4 OVM_INFO @ 150 [OUTPUT] Result: 02fa
5 OVM_INFO @ 180 [OUTPUT] A: 55  B:ff   OP:and_op
6 OVM_INFO @ 210 [OUTPUT] Result: 0055
7 OVM_INFO @ 240 [OUTPUT] A: 55  B:ff   OP:xor_op
8 OVM_INFO @ 270 [OUTPUT] Result: 00aa
9 OVM_INFO @ 340 [OUTPUT] A: 22  B:94   OP:no_op
```

FIGURE 15-12. Results of Running the Test Bench

We can use a calculator to see whether the TinyALU is working or we can write a script that reads the operations and checks the results.

For example, our first operation says

```
0xff + 0x01 = 0x0100
```

Our hex calculator says this is correct. It also agrees with this operation from lines 3–4:

```
0xfe * 0x03 = 0x02fa
```

We now have a test bench that works entirely at the transaction level. By translating to and from transactions, we were able to make a file of results that keeps us from having to look at waveforms to do our debugging.

15.8 Summary

In this chapter, we put the pieces together and created a transaction-level test bench. The test bench takes transactions and writes them to a file. Then it converts them to RTL signals and drives the DUT with them. Then, we converted the results from the DUT back to transactions and wrote them to the same file.

We were able to create a file of results that we could check by inspection or with a script. We saw that the abilities to write tests easily and review results easily are the principle advantages of a transaction-level test bench.

We can now crank through tests and write thousands of transactions. We've completed Step 4, Transactions, of the seven steps to the complete test bench.

Although creating transaction-level test benches allows us to crank out tests easily, it does create a problem when it comes to looking at the simulation output. The TinyALU is a simple design, and our technique of writing to a file and inspecting this file was sufficient.

That said, what happens if we have multiple output channels? Do we write to multiple files? How do we examine the files and make sure they are correct? The problem gets very difficult very quickly.

What we really want is a test bench that checks itself, and that's what we'll do in Step 5, Self-Checking Test Benches.

16 Creating a Transaction Predictor

IN THIS CHAPTER:

- *Creating a Transaction-Level Predictor*
- *Creating the TinyALU Predictor*
- *The* `predictor` *Module*
- *Running the Simulation*

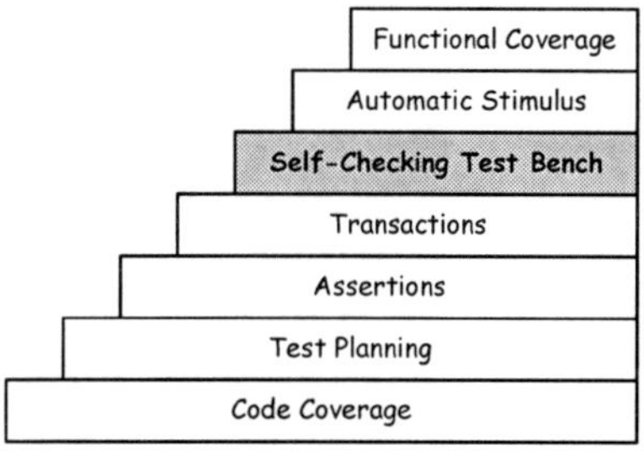

Change is a fact of design. Until we find the last bug or implement the last feature, our designs will change constantly over the course of a project. After all, when we're done changing the design we can just ship it.

The problem is that change introduces bugs. In a system as complex as a typical digital design, any change can have unintended consequences. It is the rare (and perhaps foolish?) engineer who confidently says, "This change will absolutely not create any new bugs."

This is not a new problem. In fact, software companies have dealt with bugs introduced by change for over 20 years. Their solution, and our solution, is to create a set of nightly tests that check the design at the end of the day and look for new bugs. We call this nightly simulation a *regression run,* because it is looking for places where the design and gotten new bugs in functions that previously worked fine—that is, regressed.

The self-checking test bench is a key part of this nightly regression run. If your test bench can't check itself, then you can't do a nightly regression run. You won't have time to check the results every morning, and you'll eventually become numb to the process and allow bugs to slip through. Instead, you want to kick off a self-checking regression run, go home, and come in the next morning to see if there are any bugs.

Now that we have transactions under our belt, we'll be able to do some incredible things quite easily. The first is creating self-checking regression test benches.

We saw that transaction-level test benches made it easy to write tests and even allow the test bench to generate transactions and exercise the DUT by itself. Unfortunately, we also saw that it can be difficult to tell if a design worked.

The old approach of looking at waveforms as a way of checking the behavior could never work. There is too much data in an transaction-level test bench. The best we could do is some spot checking at interesting points on the waveform, but it's unlikely that we'd see subtle bugs that could take forever to figure out in the lab.

Our test plan can help us. In that plan, we defined functionality in terms of transactions. By writing the transactions into a file, we can compare the file to the test plan and see whether we are getting the correct behavior. Once again, though, our visual inspections would need to be limited.

To avoid having to do visual inspections, we could write a script that reads the file of transactions and determines whether they are correct. However, such a script would run into trouble on a design with multiple channels. Even if we gave each channel its own file, the script would have to decide how to line the transactions up to figure out whether the design worked.

Fortunately, we have a way around this problem. It's easy to make a transaction-level test bench check itself. There are two steps:

1. Predict the output transactions on every output channel.
2. Compare them with the actual outputs from the RTL-level DUT.

All we need to do is write one or more *predictor* modules to predict which transactions should appear on the output channels. Once we have predictors, we can pair them with *comparator* modules that ensure that the transactions we get out of our RTL-level DUT are the same as the ones we get from our transaction-level predictor.

Predictor modules implement the expected behavior of a DUT at the transaction level. These modules have copies of all the input and output channels in the DUT. They receive the same input transactions as the DUT and they generate the output transactions that we expect from the DUT.

Once we have these predicted output transactions, we can compare them to the ones that come from the RTL DUT's responder modules.

16.1 Creating a Transaction-Level Predictor

To create a self-checking test bench, we need to create a model of how the DUT is supposed to behave, and then we need to check the DUT against that model. In this chapter, we'll focus on creating the model, and in the next chapter we'll show how to make the comparisons.

Transaction-level modeling hides the details of how an RTL DUT operates. This is valuable when it comes to creating a predictor. Writing a predictive model that works at the signal level would be just as much work as creating the design itself, but writing a transaction-level model is much easier.

You might be thinking that writing a predictor for your design is very difficult, but the truth is, you've already got the model. It's just that it's in your head.

When you write a test bench and look at the output, you are implementing a predictive model. You are saying, "Hmmm. Is this the right output? Yes, it is. I'm OK." In doing that, you are mentally running an algorithm that checks the design's behavior against your understanding of it.

When you write the predictor, you simply take the mental model and write it down in SystemVerilog in terms of transactions. You will have done most of the work already by creating a table in your test plan like Table 3-3 on page 32.

The key to writing a successful predictor is to keep all your thinking at the transaction level. For example, if you have a UART that has a Wishbone bus on one side and a serial port on the other, you could think about all the signals you need to drive on one side and to see driven on the other side. However, you could just say the following:

If I see a write transaction on the Wishbone side, I expect to see an output transaction with the same data on the serial side.

Remember that you already handled all the signal complexity when you wrote the driver and the responder modules. Now you can just take advantage of that work when creating the self-checking part of your test.

Let's write a transaction-level predictor for the TinyALU as an example.

16.2 Creating the TinyALU Predictor

Transaction-level predictors are really channel predictors. They take the input transactions on all the input channels and predict what will appear on the output channels.

A predictor module is a transaction-level module in that it has only `tlm_fifos` on its port list. You provide the predictor with one `tlm_fifo` per channel. The predictor reads the input channels and writes to the output channels.

Our TinyALU design has only two channels, so the predictor can fit into a test bench like this:

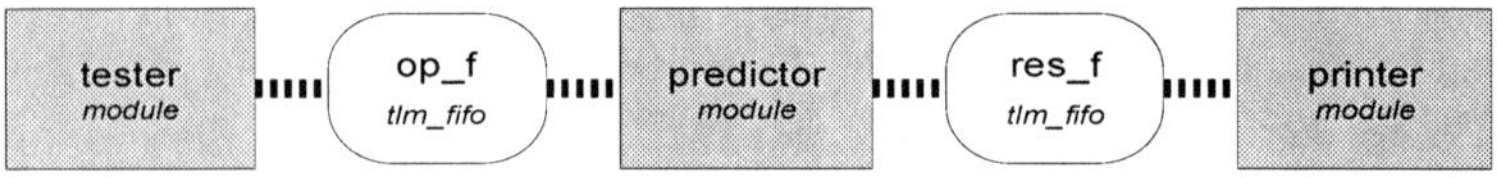

FIGURE 16-1. TinyALU Predictor in the RTL Test Bench

In this example, we have plugged our transaction-level predictor into the test bench we used in the last chapter. We've replaced `driver`, `RTLDUT`, and `responder` with a module called `predictor`. This module takes `alu_operation` transactions in from the `tester` and writes `alu_result` transactions out to the `printer`.

Our goal is to test our TinyALU predictor to make sure it runs properly. Presumably, we've had a code review to make sure that the predictor itself is working according to the specification. We could also test this predictor by using the same script we described for the RTL test bench in Chapter 15.

Here is the code for the top level:

```
16 import     ovm_pkg::*;
17 import     tinyalu_pkg::*;
18 `include   "tinyalu_defines.svh"
19
20 module top;
21    tlm_fifo #(alu_operation) op_f;
22    tlm_fifo #(alu_result)    res_f;
23
24    generator      tester    (.op_f(op_f));
25    alu_predictor  predictor (.op_f(op_f),  .res_f(res_f));
26    result_printer printer   (              .res_f(res_f));
27
28  initial begin
29       ovm_top.set_report_verbosity_level(300);
30       ovm_top.set_report_id_action(`OUTPUTID, OVM_NO_ACTION);
31       op_f   = new("op_f");
32       res_f  = new("req_f");
33       fork
34          tester.run();
35          predictor.run();
36          printer.run();
37       join_none
38    end
39 endmodule
```

FIGURE 16-2. The TinyALU TLM Test-Bench Top Level

- **Line 25**—We replaced the `driver`, `RTLDUT`, and `responder` modules with the predictor module.
- **Line 29**— We set the verbosity to 300 so we can see messages printed to the screen.
- **Line 30**—We associated the `'OUTPUTID` message id to `OVM_NO_ACTION`. Now the printer will not print to a file.

The rest of the test bench works exactly the same way and tests the `predictor` module.

16.3 The `predictor` Module

All predictor modules have the same strategy. They take transactions from input channels and create transactions on output channels. Because they are working at the transaction level, they are much easier to write than RTL models.

When we put predictors into the test bench, we feed them the same transactions we feed the DUT, and then compare their output to the output from the responder module that reads the DUT's output. If the two don't match, we have a problem.

In this example, we are writing a predictor for the TinyALU. We are going to feed the predictor `alu_operation` transactions and get back `alu_result` transactions.

We create a transaction-level predictor for the TinyALU in the code below. Notice that this version of the TinyALU has no clock. It does the following:

1. Reads an `alu_operation` transaction off the `op_f` FIFO and puts it into the variable `req`.[1]
2. Uses the `op` data member to control a case statement.
3. Creates an `alu_result` transaction by using the correct operation on the operands from the `req` transaction.
4. Puts the new `alu_result` transaction into the `res_f` FIFO.

1. Technically, we are reading a handle out of the `tlm_fifo` and putting the handle into the variable `req`. But it gets tiresome to say "getting the handle" and "putting the handle" constantly, so we take a shortcut by saying "we read a transaction and put it into `req`."

Here are the details of our TinyALU predictor:

```
16 import ovm_pkg::*;
17 import tinyalu_pkg::*;
18 `include "tinyalu_defines.svh"
19
20 module alu_predictor(ref tlm_fifo #(alu_operation) op_f,
21                      ref tlm_fifo #(alu_result)    res_f);
22
23    alu_operation req;
24    alu_result    res;
25    string        err_string,m;
26
27    task run;
28       forever begin
29          $sformat(m,"%m");
30          op_f.get(req);
31          ovm_top.ovm_report_info(m,req.convert2string(),300);
32          case (req.op)
33             add_op: res = new(req.A + req.B);
34             and_op: res = new(req.A & req.B);
35             xor_op: res = new(req.A ^ req.B);
36             mul_op: res = new(req.A * req.B);
37             no_op:  begin end
38             default: begin
39                $sformat (err_string,"BAD OP IN TLM ALU: %s",req.op);
40                ovm_top.ovm_report_fatal(m,err_string,0,`__FILE__,`__LINE__);
41             end
42          endcase
43          if (req.op != no_op) res_f.put(res);
44       end
45    endtask
46 endmodule
```

FIGURE 16-3. TinyALU Behavioral Model

- **Lines 20–21**—This model has two `tlm_fifos` on the port list. It takes in `alu_operations` and delivers `alu_results`.
- **Lines 23–24**—We define two object variables. The variables `res` and `req` hold the `alu_operation` and the `alu_result` transactions.
- **Line 27**– Here is the ubiquitous `run()` task with a `forever` loop.
- **Line 30**—Get the operation from `op_f` and put it into the `req` variable.
- **Line 31**—Print the `alu_operation` to the screen with a verbosity of 300.
- **Line 32**—This case statement decodes the operation in the transaction. It references `req.op`.
- **Lines 33–36**—We call `new()` to put a new `alu_result` object into `res`. We do the operation in this step as well by passing an expression to the `new()` method.
- **Line 37**—Ignore the `no_op`.
- **Lines 38–41**—Bail if we get a bad operation. This indicates that the `alu_operation` definition doesn't match our expectations.
- **Line 43**—We put the result transaction into `res_f`—unless it's a `no_op`.

That's all there is to this behavioral model. It is simple enough that we can verify it by inspection. In complex designs, we would use a code review or sample test runs to ensure that our model follows the specification.

16.4 Running the Simulation

We'll just print the results to the screen to make sure the simulation is working properly:

```
# ----------------------------------------------------------------
# OVM-2.0
# (C) 2007-2008 Mentor Graphics Corporation
# (C) 2007-2008 Cadence Design Systems, Inc.
# ----------------------------------------------------------------
# OVM_INFO @ 0 [top.predictor.run] A: ff  B:01   OP:add_op
# OVM_INFO @ 0 [top.predictor.run] A: fe  B:03   OP:mul_op
# OVM_INFO @ 0 [top.printer.run] Result: 0100
# OVM_INFO @ 0 [top.predictor.run] A: 55  B:ff   OP:and_op
# OVM_INFO @ 0 [top.printer.run] Result: 02fa
# OVM_INFO @ 0 [top.predictor.run] A: 55  B:ff   OP:xor_op
# OVM_INFO @ 0 [top.printer.run] Result: 0055
```

FIGURE 16-4. Running the TinyALU Predictor

The `predictor` and `printer` are both transaction-level modules, so this design runs in zero time. We can see the predictor reading transactions and writing them to the `tlm_fifo`. It doesn't suspend until it tries to write the `mul_op` transaction, and then the printer gets to work.

16.5 Summary

In this chapter we created a predictor. We saw that self-checking test benches focus on channel-level communication. They look at the transactions going into the DUT through each channel and make predictions about what transactions will come out.

This approach to self-checking relies upon the `driver` and `responder` handling the details of the DUT-level communication. Working at the transaction level allows us to create a model of our device much more easily than it would be to work at the signal level.

We created a predictor for the TinyALU, a DUT with one input channel and one output channel. In the next chapter, we'll combine our predictor with a comparator to create the self-checking test bench.

17 Creating a Self-Checking Test Bench

In This Chapter

- *Assembling a Self-Checking Test Bench*
- *Top Level of the Self-Checking Test Bench*
- *Implementing the TinyALU Comparator*

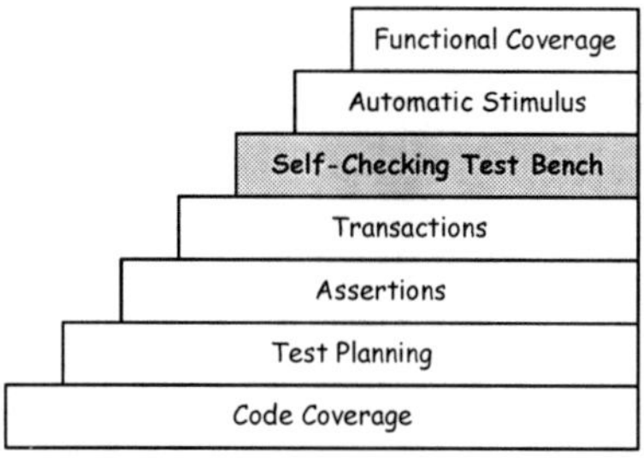

We took the first step towards a self-checking regression test bench in the preceding chapter, when we created a predictor—a transaction-level model of our design. This model ignores all the details about state machines, clocks, and pin wiggles, and simply recreates the kinds of transactions we'll see come out of our design, given the transactions that go in.

The next step is to create a *comparator*, which looks at the predicted output transactions and compares them to real output transactions.

The key to our self-checking test bench comes from the test plan. That plan decomposed our design into a set of input and output channels. We used transactions on those channels to define our chip's behavior and write a predictor. Now we'll write comparators for all the DUT's output channels to compare the predicted transactions to the real ones.

In most designs, for example, memory controllers or bus interfaces, it's easy to create a self-checking test bench this way, because the output transactions come out of the DUT in the same order they went in.

There are designs that reorder the transaction randomly—for example, arbiters. In these cases our comparator needs to store up all the transactions from the predictor and then match them against transactions that come out of the DUT. If all the transactions match, then our design works properly.

The TinyALU is the type of design where the output transactions mirror the input transactions directly.[1] Because the TinyALU has only one output channel, we are going to add one comparator to the design. In cases of designs with multiple output channels, such as the TinyCache from Chapter 2, we would have multiple comparators.

17.1 Top Level of the Self-Checking Test Bench

A self-checking test bench sends its transactions on a twisting journey. On the input side, it creates two copies of all input transactions. The test bench converts one of the copies into signals and uses them to drive the RTL DUT. It puts the other copy into a predictor. The predictor creates output transactions that are based on the DUT's expected behavior.

On the output channel, the responder converts DUT signals into an output transaction and makes a copy. It passes one copy of the output transaction to the comparator, and it passes the other copy downstream through the test bench (assuming there is a downstream module in the test bench.)

Here is the top level of the TinyALU's self-checking test bench:

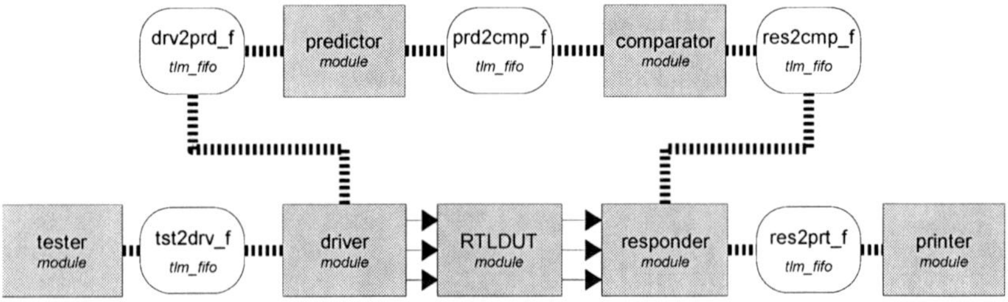

FIGURE 17-1. TinyALU Self-Checking Top Level Diagram

1. You can download examples of complex designs, as well as discuss your own designs, on www.fpgasimulation.com.

The bottom part of this diagram is the same as the test bench in Figure 15-1 on page 182. The transactions come in from the `tester` and go to the `printer`. We've added ports to the `driver` and `responder` to implement self-checking behavior on the single TinyALU output channel.

The `driver` now passes a copy of the transaction to the `predictor` through `drv2prd_f`. The `predictor` makes its prediction and passes the result to the `comparator` through `prd2cmp_f`.

Meanwhile, the responder has collected a result from the RTL DUT and created two copies of an `alu_result` transaction. One it passes downstream through the test bench, and the other it passes to the `comparator` through `res2cmp_f`.

The comparator reads the `alu_result` from `res2cmp_f` and compares it to the `alu_result` it got from the `predictor`. If they are the same, then we are OK and move on to the next transaction.

That's all there is to creating a transaction-level self-checking test bench. We've already done the hard work in the driver and responder, which hide the complexity of the signals. We just need to write a `predictor` and `comparator`.

We implement one comparator per output channel. In a design with several output channels, we have a choice of implementing one predictor per channel, or creating one big predictor that models the entire DUT. Then we would take the output transactions from that predictor and send them to comparators for each of the output channels.

17.1.1 The `tlm_fifo` Naming Convention

Notice that we've added a new naming convention to our `tlm_fifo` objects. The `tlm_fifos` are actually bidirectional. Because either module can call the `put()` method and put transactions into the `tlm_fifo`, we'll create names for the `tlm_fifos` that remind us of where they are going.

For example, we have a `tlm_fifo` that goes from the `driver` to the `predictor`, so we call it `drv2pred_f`. We also have a connection between the predictor and the comparator, so we call that one `prd2cmp_f`.

17.1.2 Implementing the Top Level in Code

The top-level code for the self-checking test bench is roughly the same as the code for the RTL test bench. The biggest difference is that we don't create an output file and we don't write transactions to it. Instead, we rely upon the self-checking nature of the test bench to tell us whether our design worked.

Here's the code:

```
module top;
    tlm_fifo #(alu_operation) tst2drv_f;
    tlm_fifo #(alu_result)    res2prt_f;
    tlm_fifo #(alu_operation) drv2prd_f;
    tlm_fifo #(alu_result)    prd2cmp_f;
    tlm_fifo #(alu_result)    res2cmp_f;

    wire clk, reset_n, done, start;
    wire [2:0] op;
    wire [7:0] A,B;
    wire [15:0] result;

    generator      tester      (.op_f(tst2drv_f));

    alu_driver driver
      (.op_f(tst2drv_f),.pred_f(drv2prd_f),
       .clk(clk), .reset_n(reset_n),.start(start),.done (done),
       .op(op),.A(A),.B(B));

    tinyalu RTLDUT
      (.clk(clk),.reset_n(reset_n),.start(start),.op(op),.A(A),.B(B),
       .done (done),.result(result));

    alu_responder responder
      (.res_f(res2prt_f),.comp_f(res2cmp_f),
       .clk(clk),.reset_n(reset_n), .done(done),.result(result));

    result_printer printer   (.res_f(res2prt_f));

    alu_predictor  predictor (.op_f  (drv2prd_f), .res_f(prd2cmp_f));

    alu_comparator comparator(.predicted_f(prd2cmp_f),.actual_f(res2cmp_f));

  initial begin
       ovm_top.set_report_verbosity_level(200);
       ovm_top.set_report_max_quit_count(2);
       ovm_top.set_report_id_action(`OUTPUTID, OVM_NO_ACTION);

       tst2drv_f  = new("tst2drv_f");
       res2prt_f  = new("res2prt_f");
       drv2prd_f  = new("drv2prd_f");
       prd2cmp_f  = new("prd2cmp_f");
       res2cmp_f  = new("res2cmp_f");

       fork
          tester.run();
          predictor.run();
          comparator.run();
          printer.run();
       join_none
    end
endmodule
```

FIGURE 17-2. The TinyALU Self-Checking Top Level

- **Lines 21–25**—Declare the handles for all our `tlm_fifos`.
- **Lines 27–30**—Declare the wires for all our RTL-level modules.
- **Lines 32–51**—Instantiate the test bench modules that drive the DUT and check the results.
- **Line 35**—We've added `drv2prd_f` to the driver port list, to send a copy of the incoming transactions to the `predictor`.
- **Line 44**—We've added `res2cmp_f` to the `responder` port list, to send a copy of outgoing transactions to the `comparator`.
- **Line 49**—Instantiate the `predictor`. Connect it to the `driver` and `comparator`.
- **Line 51**—Instantiate the `comparator`. Connect it to the `predictor` and the `responder`.
- **Line 54**—Set the verbosity to 200, so we filter out informational messages and see only error messages.
- **Line 55**—Quit after two errors.
- **Line 56**—Turn off actions related to the `` `OUTPUTID`` message ID. We are no longer writing to a file.
- **Lines 65–68**—Launch threads for all the `run()` tasks in our transaction-level modules.

The attentive reader may wonder, "Why do we have a `printer` module if it's not printing?"

Good question. We could have just taken the output from the `responder` and put it right into the `comparator`. In fact, in cases like this, where there is only an input/output relationship, this is the most efficient solution to the problem.

On the other hand, there are many designs where we have to model downstream logic. For example, the TinyCache that we discussed in Chapter 2 has a downstream interface to the main memory. In that case, we want to siphon transactions off rather than just run them into a comparator.

17.1.3 Running the Test Bench

I've introduced an error in the predictor in this example, so that we can see the test bench catch errors.[2] Here's the output:

```
# ----------------------------------------------------------------
# OVM-2.0
# (C) 2007-2008 Mentor Graphics Corporation
# (C) 2007-2008 Cadence Design Systems, Inc.
# ----------------------------------------------------------------
# OVM_ERROR @ 270 [top.comparator.run]
#       Golden Result: Result: 00ff
#       Actual Result: Result: 00aa
# OVM_ERROR @ 370 [top.comparator.run]
#       Golden Result: Result: 00f2
#       Actual Result: Result: 0032
#
# --- OVM Report Summary ---
#
# Quit count reached!
# Quit count :          2 of          2
# ** Report counts by severity
# OVM_INFO :    0
# OVM_WARNING :    0
# OVM_ERROR :    2
# OVM_FATAL :    0
# ** Report counts by id
# [top.comparator.run  ]     2
# ** Note: $finish     : ../../ovm/src/base/ovm_report_object.svh(149)
```

FIGURE 17-3. Catching Errors

In this example, there is a mismatch between the predicted result and the actual result, so we see errors. We had set up the messaging system to quit after two errors (on line 55 in Figure 17-2 on page 212.)

In a larger design we may have tens or hundreds of tests running. In those cases we would have added logging behavior to our error messages. We would then look for error files with data in them, to see if we had an error anywhere in our hundreds of tests. Alternatively, we could log all errors to the same file and then check the file at the end of simulation.

2. The code is on www.fpgasimulation.com, and finding the error has been left as an exercise for the reader.

17.2 Implementing the TinyALU Comparator

The comparator module is the key to the self-checking test bench. It takes a transaction from the `predictor` and compares it to the output of the `RTLDUT`. It uses the `comp()` method in the `alu_result` transaction to do this.

The `comp()` method answers the question, "Is the data in that transaction the same as the data in this transaction?" In the case of the `alu_result`, this is a simple question to answer. We just confirm whether the `result` data member is the same in both. Here is the `comp()` method for the `alu_result` transaction as defined in `tinyalu_pkg.sv`:

```
function bit comp (alu_result r);
  return (r.result == result);
endfunction
```

FIGURE 17-4. The `comp()` Method for `alu_result`

It hardly seems worth the trouble to create an entire method to compare a single data member. In the case of the TinyALU that is probably true. However, by getting into the habit of creating a `comp()` method, we set ourselves up to handle complex transactions.

For example, what if our data were a floating-point number, and our predictor read numbers from a table generated by a software program? Would these floating-point numbers match up exactly, or would there be a margin of error to some degree of precision? In those cases, the question of whether two transactions are the same would require some calculation in the predictor. We would capture that calculation in the `comp()` method.

The TinyALU comparator is very simple. There is a one-to-one ratio between input transactions and output transactions, so every time we read a transaction from `RTLDUT`, there should be a corresponding transaction from the predictor. This gives us a simple algorithm:

1. Read a transaction from `actual_f` (actual result).
2. Read a transaction from `predicted_f` (predicted result).
3. Compare the transactions.

We read the `actual_f tlm_fifo` first, because the predictor runs in zero time while the DUT takes at least one clock cycle. Therefore, the predicted value will always be ready by the time the actual value is ready.

Here's the code:

```
 4  module alu_comparator(
 5       ref tlm_fifo #(alu_result) predicted_f,
 6       ref tlm_fifo #(alu_result) actual_f);
 7
 8       alu_result predicted, actual;
 9       string     m,message;
10
11       task run();
12         $sformat(m,"%m");
13         forever
14           begin
15             actual_f.get(actual);
16             predicted_f.get(predicted);
17             $sformat(message,"\n\tGolden Result: %s\n\tActual Result: %s",
18                        predicted.convert2string,actual.convert2string);
19             if (predicted.comp(actual))
20               ovm_top.ovm_report_info (m,message,300);
21             else
22               ovm_top.ovm_report_error(m,message,200);
23           end
24       endtask
25  endmodule
```

FIGURE 17-5. TinyALU `comparator` Module

- **Lines 5–6**—The comparator reads two `alu_result` transactions, one from the `predictor` and one from the RTL.
- **Line 15**—Read the actual transaction from the RTL.
- **Line 16**—Read the predicted transaction from the predictor.[3]
- **Lines 17–18**—Create a string that contains both transactions.
- **Line 19**—The comparison!
- **Line 20**—Print out an info message if we pass with a verbosity of 300.
- **Line 22**—Print out an error message if we fail with a verbosity of 200.

Our test bench can now check itself.

3. You could also use the `try_get()` method here within an `assert` statement, and a call to `ovm_report_fatal()` if the `assert` fails. This would kill the simulation if the prediction were not ready—a sure sign of a bug in the test bench.

17.3 Summary

In this chapter we completed Step 5, The Self-Checking Test Bench. Now our test bench will tell us if it finds an error.

We examined the top level of the self-checking test bench and saw that it adds one comparator for each output channel in our DUT. The comparator takes predictions from the predictor and compares them with the actual RTL output.

Now that our test bench can check its own output, we are ready for the next, and most powerful, part of transaction-level test benches. We are ready to let the test bench write its own tests. We'll do that in Step Six: Automatic Stimulus.

18 Introduction to Automatic Stimulus

IN THIS CHAPTER:

- *Constrained Random Testing*

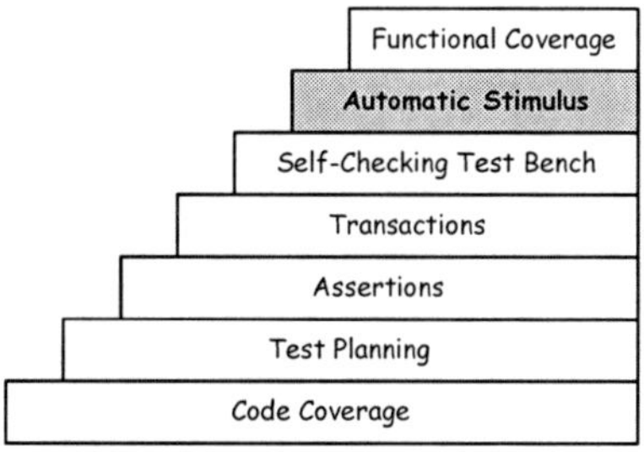

In Chapter 16, Creating a Transaction Predictor, I said that we could run tests all night long and then view the results the next morning. This raises the question, "Who's going to write tests that run all night long? Are you going to do it, Salemi? Because I'm not."

This is an important question. Writing tests can take hours or days. Some ASIC design teams have twice as many verification engineers as design engineers. The enormous number of tests is a function of how quickly the scope of digital functionality explodes.

All digital systems are gigantic state machines. Our verification job is to make sure that all the possible states in these machines work the way they're expected to work. We call the complete collection of all possible functionality the *state space,* and it is the product of all the possible places our digital design can go.

At the grossest level, we can determine our state space during our test planning phase. We multiply all the things our design can do by the different ways it can do them, and that gives us the number of things we need to test.

For example, www.opencores.org[1] defines an open-source bus for interblock communication called the *Wishbone bus*. The Wishbone bus allows users to do read, write, and read-modify-write operations. It also allows operations to be different widths, and it allows block operations for read and write.

If we are testing a Wishbone interface we want to make sure we cover every possible behavior. One way to do that is to create a table of the functionality where we calculate how many tests we'll need to test all the operations in all the addressing modes given the functionality we implement:

TABLE 18-1. The Number of Tests Needed to Test the Wishbone Bus

	8-Bit	16-Bit	32-Bit	64-Bit
Read	1	2	3	4
Write	2	4	6	8
Block Read	3	6	9	12
Block Write	4	8	12	16
Read-Modify-Write	5	10	15	20

If our design does only eight-bit reads, we need one test. If it does eight-bit reads and writes we need two tests. If we implement the Wishbone specification fully, we need 20 tests for one Wishbone block to try every combination of operations and addressing modes.

Of course, it doesn't end there, because this is only counting unidirectional communication. If we have two Wishbone blocks talking to each other, we need to make sure they try talking in both directions. That's 40 tests. If we have three blocks we, now have an n^2-n situation. This is the number of possible communication paths between n blocks. Three blocks have six (3^2-3) ways of talking, so that's `20*(9-3)=120` tests.

Of course, none of this checks what happens when the blocks try to arbitrate for who gets the bus. Nor does it test the order of the operations between the blocks. Can we do a read followed by a read-modify-write? What if one block does a read and the other block does a write back to the same address? What if we run the same operation twice in a row?

1. http://www.opencores.org/projects.cgi/web/wishbone/wishbone

Also, none of these tests have anything to do with what the blocks actually do. We're just testing whether they can communicate. What about their base functionality? And what happens if one of these blocks bridges to a PCI-Express bus? That entails an entirely new set of tests.

The state space of even relatively simple digital designs explodes, and with it our hope of trying to write a test for all the combinations of operations that could take place in the design.

This strategy of trying to write a test for each behavior in the design is called *directed test,* and it quickly runs out of steam. Not only does the state space get out of hand, but we can't think of everything. If it never occurred to us that a certain killer combination of events could hurt us, how can we write a test for it?

The key is to simulate what goes on in the real world—i.e., the lab. The lab quickly finds bugs even as it makes it difficult to debug them. This is because a lab environment throws random transactions at our block in ways we hadn't considered. This approach quickly explores the entire state space and finds the states where our design doesn't work. We want to create a similar situation in our simulated test bench.

18.1 Constrained Random Stimulus

As an applications engineer, I have often been asked, "Do you have a tool that can write tests for me?" This is a little like asking, "Do you have a word processor that can write novels for me?" The short answer is, "No."

However, what if we asked a different question? What if we asked, "Do you have a tool that can write haiku?" We might just be able to answer that question with "Yes."

Haiku is a simple three-line poetry form with five syllables on the first line, seven on the second line, and five on the third line. For example:

My project is doomed.
No one verified the chip.
And now: The Deathmarch.

Could we write a Perl script that would generate haiku? Easily. We'd create lists of words with different numbers of syllables and generate them randomly until they matched the syllable count. It might not be readable haiku, but it would match the 5-7-5 pattern. If we got fancier we could require a noun-verb-noun structure of some sort and get readable sentences.

The same randomizing approach allows us to create a test bench that writes its own tests. We've created a transaction-based test bench that accepts complete packets of stimuli (transactions) and converts them to signals. By creating those transactions randomly, we can create tests that push our design to unexpected places.

We've already had a taste of randomized stimuli in the TinyALU tester. The tester creates ten random transactions and sends them into the test bench, where the DUT operates on them and the scoreboard checks the results. Here's the snippet of code:

```
      for (int i = 1; i<=10; i++) begin
         assert(op.randomize());
         op_f.put(op.clonecast());
      end
```

FIGURE 18-1. Ten Random Transactions

This code randomizes the `op` transaction and then puts a clone of it into the test bench through the `op_f` FIFO. These were unconstrained random transactions that could take on any legal value for eight bits or the operation enumerated type. If we had made this loop ten million times instead of ten, we would have a righteous test!

This unconstrained randomization works because the TinyALU is, well, tiny. All the possible inputs are legal and interesting. Unconstrained randomization won't work in a real system. A specification may have legal combinations of operations and operands, or there may be limits to the size of data. Having a 32-bit memory space doesn't mean that you'll be addressing 4 GB of memory. There may be blank spots in the memory space.

Also, random data may make it difficult to get to interesting cases. If we are testing a cache, we can't choose 32-bit addresses randomly and throw them at the cache—the odds of a cache hit would be too small. We need to limit the scope of our addresses to make hits more likely.

We also need a way to create random transactions that are legal for our device and meaningful for our tests. These are called *constrained random transactions* and you'll hear people use the phrase *constrained random stimulus* when they are talking about test benches that use these kinds of transactions.

SystemVerilog has language constructs that allow us to define randomizable transactions and constrain their behavior. In the next three chapters we explore these constructs in the following order:

1. **Randomization mechanics**—We'll see how to create randomizable transactions and various approaches to randomizing and constraining them.
2. **Constraints**—We'll walk through the various random constraints and see how they work with the `randomize()` method.
3. **Extending classes with constraints**—We'll show how to embed constraints into classes to control their response to the `randomize()` method.

By the end of Step 6, Automatic Stimulus, you'll be able to create randomized tests that run all night long, allowing your self-checking test bench to catch bugs while you sleep.

In the next chapter, we'll take the first step by learning how to create random numbers in Verilog.

19 Random Numbers in Verilog '95

IN THIS CHAPTER:

- `$random`—*The Mother of All Randomization*
- *Truncating and Extending* `$random`
- *Advanced Random-Number Generation*

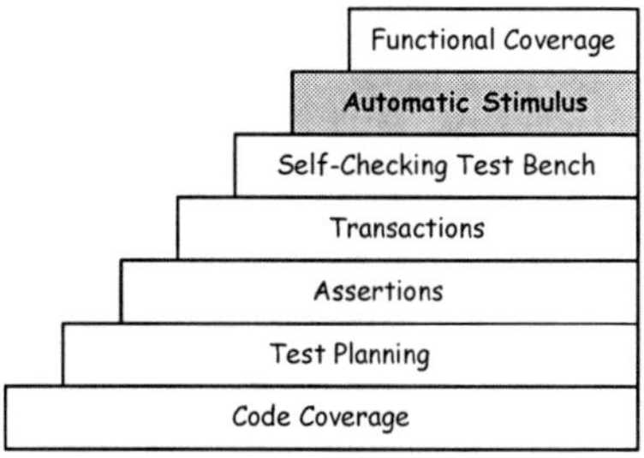

Now we'll discuss techniques that are used by the best digital-design groups in the world. These engineers, from companies such as Apple, Nvidia, and Cisco cannot afford to bring an unverified chip to the lab, because each chip version can cost upward of a million dollars. The chips need to arrive from the vendor and work as soon as they are plugged into the board. Anything less is failure.

Confronted with such dire circumstances, these engineers don't even try to write directed tests for all the possible chip behaviors. Instead, they use constrained random stimulus to generate their tests. They easily create different tests by changing the constraints.

Once you start creating test benches that generate their own stimuli, you won't want to go back to writing directed tests. Instead, you'll reserve your directed tests for corner cases in your code coverage, where you want to force the design into unusual behaviors.

In this chapter, we'll look at randomization tools available in Verilog. These provide some simple ways of quickly creating a random number. In our next chapter, we'll look at additional randomization tools available in SystemVerilog.

19.1 `$random`—The Mother of all Randomization

While SystemVerilog has powerful randomization features, the original Verilog language had its own basic randomization tool, called `$random`. This is a great tool for those times when you simply want some random bits. It also has the advantage of not needing SystemVerilog, so it works with simple Verilog simulators.[1]

The definition for `$random` looks like this:

```
integer $random[inout (seed)]
```

`$random` accepts an optional variable as a seed and then returns a 32-bit signed number in response. `$random` changes the seed each time it's called, so that it delivers a different random number each time. Calling `$random` with the same seed value always returns the same result.

Here is an example of creating ten random numbers with `$random`:

```
module top;

integer my_num, seed, i;

initial
   begin
      $display ("-- No Seed --");
      for (i = 1; i <= 5; i=i+1) begin
         my_num = $random;
         $display(my_num);
      end
      $display("-- Seed initially set to 10 --");
      seed = 10;
      for (i = 1; i <= 5; i=i+1) begin
         my_num = $random(seed);
         $display("Random Numb: %0d   New seed: %0d",my_num,seed);
      end
   end
endmodule
```

FIGURE 19-1. Simple Randomization with Integers

1. There are random-number generator packages written for VHDL, but these are not built into the language. Several such packages appear when you search for "VHDL Random Number Generation" in Google.

- **Line 9**—This demonstrates calling `$random` without a seed. This is the same thing as calling `$random` with a seed whose initial value is zero.
- **Line 15**—This is how we call `$random` with a seed variable. `$random` changes the value of `seed` on each call. If you were to pass just a constant to `$random` instead of a variable, you would get the same "random" number each time you called `$random`.

The results are random-looking, as expected:

```
# -- No Seed --
#   303379748
# -1064739199
# -2071669239
# -1309649309
#   112818957
# -- Seed initially set to 10 --
# Random Numb: -2146792448   New seed: 690691
# Random Numb: -1686787018   New seed: 460696424
# Random Numb: 576097604   New seed: -1571386807
# Random Numb: 1855681501   New seed: -291802762
# Random Numb: -390931759   New seed: 1756551551
```

FIGURE 19-2. Random Results from `$random`

Notice that `$random` returns a signed integer. Verilog uses a two's complement to represent negative numbers, so the negative numbers represent cases where the most significant bit is `1`. These numbers are printed as negative numbers because we've declared `my_num` as an integer. However, we don't work with integers in hardware, but rather we work with bits, and that's the next subject.

19.1.1 Truncating and Sign-Extending `$random`

One of the nice (or dangerous) things about Verilog is that it automatically manages truncation and sign-extension when two operands have different lengths. We use this capability when we truncate or extend the numbers from `$random`. There are three rules that manage truncating and extending:

1. If the target is shorter than the source, then we truncate all the bits above the width of the target.
2. If the target is longer than the source and the source is an unsigned number, then we pad the extra bits with zero.
3. If the target is longer than the source and the source is a signed integer, then we copy the most significant bit of the target into the extra bits in the source.

By letting Verilog manage our truncation and sign-extension, we can fit random numbers into any length of register. Here's an example. This module creates a random number and then places it into three target variables. One of these is 8-bits wide, and the other two are 64-bits wide:

```
1  module top;
2
3     reg [7:0]  reg8;
4     reg [63:0] reg64_ext, reg64_zero;
5     integer    my_random, i, seed;
6
7     initial
8        begin
9           seed = 5;
10          for (i = 1; i<= 4; i=i+1) begin
11             my_random = $random (seed);
12             reg8       =  my_random;
13             reg64_ext  =  my_random;
14             reg64_zero = {my_random};
15             $display;
16             $display ("my_random(2's comp):  %d", my_random);
17             $display ("my_random(hex      ):  %h",my_random);
18             $display ("reg8:                  %h", reg8);
19             $display ("reg64_ext:             %h", reg64_ext);
20             $display ("reg64_zero:            %h",reg64_zero);
21          end
22       end
23 endmodule
```

FIGURE 19-3. Truncating and Extending `$random`

- **Line 9**—Use a seed for this program to control the random numbers The number 5 gives us the desired result—two negative random numbers and two positive random numbers. Verilog replicates this result from run to run because of the seed.
- **Line 11**—Generate a random number and put it into an integer. This is a signed 32-bit integer. The most significant bit is 1 if the number is negative, and the number is represented as a two's complement.
- **Line 12**—Put this 32-bit number into an 8-bit register.
- **Line 13**—Put this 32-bit number into a 64-bit register.
- **Line 14**—Here is a trick that gets rid of the sign-extension. We concatenate the random number with nothing because the output of a concatenation operator is an unsigned number. This changes the random number from a signed 32-bit number to an unsigned 32-bit number. Then we put this unsigned 32-bit number into a 64-bit register and the upper 32 bits are zero.
- **Lines 16–20**—Print the numbers out, first in two's complement decimal and then and then in hex. Then print the truncated and extended results.

When I run this program I can see how the truncation and sign-extension works:

```
# my_random(2's comp):   -2147138048
# my_random(hex      ):  80054600
# reg8:                  00
# reg64_ext:             ffffffff80054600
# reg64_zero:            0000000080054600
#
# my_random(2's comp):     230383387
# my_random(hex      ):  0dbb5f1b
# reg8:                  1b
# reg64_ext:             000000000dbb5f1b
# reg64_zero:            000000000dbb5f1b
#
# my_random(2's comp):    -547878722
# my_random(hex      ):  df5808be
# reg8:                  be
# reg64_ext:             ffffffffdf5808be
# reg64_zero:            00000000df5808be
#
# my_random(2's comp):    1492801457
# my_random(hex      ):  58fa57b1
# reg8:                  b1
# reg64_ext:             0000000058fa57b1
# reg64_zero:            0000000058fa57b1
```

FIGURE 19-4. Results of Truncation and Sign-Extension

The first number demonstrates how negative numbers work in Verilog. I've generated a negative number. You can see this in the hexadecimal view of the number: `32'h80054600`. The most-significant hex number is `4'h8`, which translates to `4'b1000`, so the sign bit is set and this is a negative number.

This has no bearing when I copy this number into an 8-bit register. I take the bottom eight bits (all zeros in this case) and they go into the 8-bit register.

Things get more interesting when I copy this 32-bit signed integer into a 64-bit variable, because this is a negative number, Verilog sign-extends the most-significant bit (MSB) of the random number, and we get `f`'s in the upper 32 bits of the new number.

Then I copy the random number again, but I convert it from a 32-bit signed integer to a 32-bit unsigned integer with the `{}` operator. Now Verilog fills zeros into the upper 32 bits, because we are not copying a signed number.

You can see that all this question of sign-extension is irrelevant if we generate a positive number randomly. In that case the zeros get padded, whether or not we use the concatenation trick.

We can use the approaches in this example as long as we want to do one of two things:

1. Create a random number that is shorter than 32 bits.
2. Place a 32-bit random number into a larger variable without changing its value.

However, if we want to create a random number larger than 32 bits, we need to go back to our concatenation operator. We can create the larger number if we concatenate `$random` calls together.

For example, let's create a 50-bit random number:

```
module top;

   reg [49:0] reg50;

   initial begin
      reg50 = {$random, $random};
      $display ("reg50: %h",reg50);
   end
endmodule
```

FIGURE 19-5. Creating 50 Random Bits

This code generates 64 random bits and puts them into a 50-bit variable. Truncation takes care of the rest.

19.1.2 Advanced Random-Number Generation

The `$random` system function solves many random-number issues. However, it cannot solve all of them because `$random` delivers random numbers with a uniform distribution. The odds of generating any 32-bit number are the same as the odds of generating any other 32-bit number.

For example, what if you wanted to test a cache and you wanted your random numbers to focus around a mean and have a small chance of being far away from that mean? This is not a uniform distribution, but rather it is a normal distribution, and you can generate it with the advanced randomization functions in Verilog. There are seven functions.

TABLE 19-1. Advanced Random-Number Functions

Function Calls
`$dist_uniform(seed, start, end)`
`$dist_normal(seed, mean, standard_deviation)`
`$dist_exponential(seed, mean)`
`$dist_poisson(seed, mean)`
`$dist_chi_square(seed, degree_of_freedom)`
`$dist_t(seed, degree_of_freedom)`
`$dist_erlang(seed, k_stage, mean)`

The names of these functions represent the shape of the distribution curve you'd see if you generated a large number of functions and made a histogram. Many of these curves are quite esoteric and beyond the scope of discussion in this book. (they are discussed in detail on Wikipedia.)

There are three curves with obvious relationships to verification:

- **Uniform**—This is the most common randomization function. Each number in the range has an equal chance of being generated.
- **Normal**—This is the famous bell curve. You control the center of the bell curve with the mean and the width with the standard deviation. You can get numbers far outside the standard deviation, but the odds are low.
- **Poisson**—Poisson numbers are used in performance analysis. These numbers represent the average time between service requests to a system. For example, if I told you that people arrive to use an ATM every 120 seconds on average, you would use a Poisson distribution to create a list of wait times to simulate the ATM performance. Sometimes someone won't arrive for five minutes, but then another time two people might arrive within a second of each other.

All these function calls take a seed and use it the same way as `$random`.

19.2 Summary

Good old Verilog '95 has some simple tools for creating random numbers. We can manipulate these random numbers with the natural truncation, sign-extension, and concatenation operations in Verilog.

These tools can help us randomize transactions. We could, for example, use a `$random` function in a transaction constructor to randomize the variables; but this would be a waste of time and effort, because SystemVerilog has much more powerful and controllable randomization tools for objects. We'll examine these in the next chapter.

20 Randomizing Objects in SystemVerilog

IN THIS CHAPTER:

- *The* `randomize()` *method*
- *Declarations with* `rand` *and* `randc`
- *Modifying* `randomize()` *behavior*
- *Seeding Randomization*
- *Turning Randomization On and Off*

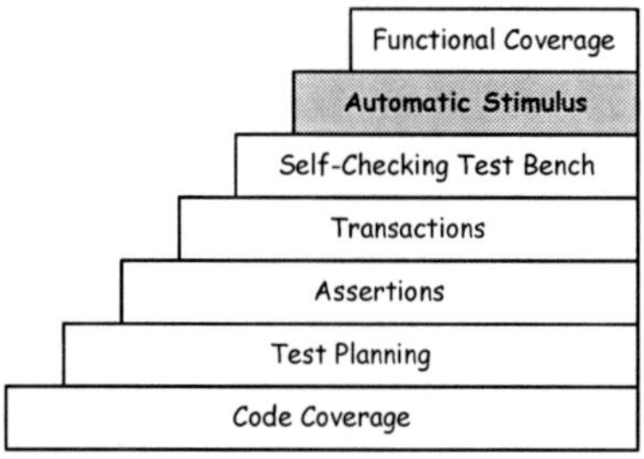

Our goal is to create a test bench that writes its own tests—or, more specifically, creates its own stimuli. This has become much easier now that we've moved to a transaction-level test bench and away from a traditional test bench, where we have to worry about which bits to toggle. Our test bench now uses objects called *transactions* to drive the DUT, and these transactions are the key to creating automatic stimuli.

Our strategy is simple. Our test bench creates transactions that contain random data and puts those transactions into the test bench. Our test bench drives the DUT and checks the results. It can run all night long and notify us if it sees a case where the predicted results don't match the actual DUT results.

However, purely random data can not do the job for us. In a complicated system, many random transactions are be illegal. For example, a memory space may contain a ROM. It would be illegal to create a write operation for that area in the memory.

Also, we may find that using purely random data doesn't match real behavior, and so it doesn't match our testing needs. Let's say we are testing a cache with a large memory space. The odds that we'll read from the same location twice are pretty low if we give all addresses the same chance of being chosen. This would give us a test that generates only cache misses.

We need to put random data into transactions, so we can test our design without having to write tests, but we don't want the data to be too random. We need to constrain the randomizer to give us the right kind of random data.

That kind of verification is called *constrained random* verification in the industry. There has been some confusion in the FPGA design community about this phrase, and that's why this book has chosen the phrase *automatic stimulus*.

The confusion lies in the use of the phrase *random testing*. This implies, to some people, a QA methodology where you choose parts randomly out of a bin and test them. This is *not* what we're talking about here. Using randomized transactions to generate automatic stimuli allows us to get to complete code coverage faster than we could by writing directed tests. It creates *more* coverage of the design, not less. With some complicated and very stateful designs, generating automatic stimuli is the only way to catch all the possible corner cases.

SystemVerilog provides powerful tools that support automatic stimulus. These tools allow us to randomize transactions within constraints and thus create useful stimuli. We are going to look at all these tools in this chapter and in the next five.

By the end of our discussion, you'll be able to create a test bench that can write its own tests.

20.1 The `randomize()` Method

The transaction object is the key to automatic stimulus. Transactions control the drivers and the drivers drive our DUT. We create automatic stimulus by creating random transactions and then feeding them to the driver.

SystemVerilog's randomization tools are much more advanced than Verilog's. In Verilog, we simply created a string of random bits and shoved them into a variable, extending or truncating the bit string as we went. With SystemVerilog, we'll be able to control our random numbers very tightly.

SystemVerilog provides inherent methods to every class, and `randomize()` is one of those methods. This method inserts random numbers into the data members that we declare as random variables.

SystemVerilog defines `randomize()` like this:

```
function int randomize();
```

This method declaration says that `randomize()` takes no arguments and returns an integer. If we successfully randomized all our numbers, then `randomize()` returns a `1`, otherwise a `0`.

We call randomize the same way we call other methods, by using the dot `(.)` operator. Here's an example from our TinyALU test bench[1] that creates ten randomized transactions:

```
32        op = new();
33        for (int i = 1; i<=10; i++) begin
34           assert(op.randomize());
35           op_f.put(op.clone());
36        end
```

FIGURE 20-1. Using the randomize() Method

- **Line 17**—Create a new transaction called `op`.
- **Line 18**—Start a for loop to randomize `op` ten times.
- **Line 19**—Randomize `op`. Notice the method call with the dot operator. The assert statement does automatic error checking.
- **Line 20**—Put a clone of `op` into the test bench.

The `randomize()` method does not automatically randomize all the variables in a class. We need to tell SystemVerilog which data members we want randomized. We do that with the `rand` and `randc` keywords.

20.2 Declarations with `rand` and `randc`

When you define a class you can declare some or all of its data members to be random variables. These are the only variables affected by the `randomize()` method.

1. From Figure 15-2 on page 183

Here's a simple example with three random variables:

```
1 class my_rand_obj;
2     rand  logic [2:0] random;
3     randc bit   [2:0] cyclic;
4     randc enum {DOGS, CATS, TREES, FLOWERS} life;
5 endclass
```

FIGURE 20-2. Defining Random Variables

When we call the `randomize()` method in `my_rand_obj`, the data members `random` and `cyclic` receive random numbers. Other than that, they act like any other variables. The `rand` and `randc` qualifiers have different behaviors.

20.2.1 The `rand` Qualifier

If you add `rand` to a data element as on line 2 in Figure 20-2, you are telling `randomize()` to choose random numbers for that data member with a uniform distribution. This means that `randomize()` can choose any number with equal probability. In the case of `random`, which is a three-bit number, `randomize()` can choose any number from `0` to `7`.

You can even use `rand` with an array. If you randomize a fixed array (where the number of elements in the array is defined at declaration), then `randomize()` puts a random number into every element in the array.

Dynamic arrays and associative arrays are different. These arrays have no fixed size, and you can add data to them dynamically to make them bigger. In these cases, you need to put data into the array before you call `randomize()`. Then `randomize()` replaces the values in the arrays with random values.[2]

20.2.2 The `randc` Qualifier

The "C" in `randc` stands for cyclic. Cyclic random variables act like a deck of cards. If you shuffle a single deck of cards and deal a Queen of Hearts off the top, you know that you won't see another Queen of Hearts until you reshuffle the

2. We'll see in a later chapter that you can use constraints to get SystemVerilog to resize a dynamic array.

deck. That is cyclic randomization. Cyclic random variables do the same thing as a deck of cards. They take on all possible values before they repeat themselves.

The `randc` qualifier allows us to create more efficient test benches. For example, if we have ten possible instructions in a processor and three addressing modes, we could allow uniform randomization to try to hit all the possible combinations (all 300 of them.) However, it could take quite a while to touch all the cases because of repeats.

If, instead, we cyclically randomize the instructions and addressing modes, we know that we'll touch them all in exactly 300 transactions. We won't see repeated combinations.

We declare variables of both types when we define a class. Each random variable at the top of the class gets its own `rand` or `randc` qualifier.

Here is an example of randomizing data in a class:

```
1  class my_rand_obj;
2     rand  logic [2:0] random;
3     randc bit   [2:0] cyclic;
4     randc enum {DOGS, CATS, TREES, FLOWERS} life;
5  endclass
6
7
8  program top;
9
10    my_rand_obj r;
11
12   initial
13    begin
14       r = new();
15       $display;
16       $write("RAND  RANDOM: ");
17       for (int i=0; i<=15; i++) begin
18           assert(r.randomize());
19           $write("%0d ",r.random);
20           if (i == 7) $write ("| ");
21       end
22       $display;
23       $write("RANDC CYCLIC: ");
24       for (int i=0; i<=15; i++) begin
25           assert(r.randomize());
26           $write("%0d ",r.cyclic);
27           if (i == 7) $write ("| ");
28       end
29       $display;
30       $write("RANDC LIFE: ");
31       for (int i=0; i<=7; i++) begin
32           assert(r.randomize());
33           $write("%0s ",r.life);
34           if (i == 3) $write ("| ");
35       end
36       $display;
37    end
38 endprogram
39
```

FIGURE 20-3. Randomizing Variables

- **Line 2**—The three-bit logic variable `random` has been qualified with the `rand` keyword. It can take on any of eight possible values regardless of its last value.
- **Line 3**—The cyclic variable is an array of three bits. Notice that the dimension comes before the variable name. This is a SystemVerilog *packed array*. Verilog '95 calls this a *vector*, as does VHDL.

  If the dimensions came after the variable, we would be describing an array of single bits and the `randc` qualifier could not be used.
- **Line 4**—We've created an enumeration of four things that are alive. The `randc` qualifier means that we'll cycle randomly through all of them before they repeat.

When we run this program we can see our random variables in action:

```
# RAND  RANDOM: 1 4 1 7 6 3 4 3 | 5 6 4 4 1 7 0 3
# RANDC CYCLIC: 0 6 2 5 1 7 3 4 | 7 6 3 5 2 4 1 0
# RANDC LIFE: CATS TREES FLOWERS DOGS | CATS DOGS FLOWERS TREES
```

FIGURE 20-4. Random Variables in Action

We see that the `rand` variable `random` can be any of the eight possible values. This is like rolling an eight-sided die. Each number is independent of the previous one. The `randc` bit variable `cyclic` can also be eight values, but it cycles through the values so that you use up all eight values before it starts over. The enumerated `randc` variable `life` also cycles through its four possible values before starting over.

20.3 Modifying randomize() Behavior

In certain cases, we may want to modify the way an object randomizes itself. Sometimes we might want to set up data in the object before it randomizes, but more often we want to look at the randomized data and modify it in some way after it's been generated.

SystemVerilog allows us to do this by defining two methods that control randomization. We can modify randomization by adding one or both of these methods to our object. The methods are `pre_randomize()` and `post_randomize()` and they look like this:

```
function void pre_randomize();

function void post_randomize();
```

The `randomize()` method calls these functions before and after it randomizes the variables. If we don't define these functions, then SystemVerilog leaves them empty and they don't do anything. If we do define them, we can change the way we randomize our object.

For example, let's say we have a transaction that requires even parity on its data. We can't randomize numbers so that they automatically have even parity[3], so we need to add the parity bit after we've created the random number. We do that by defining the `post_randomize()` method in our transaction class.

The example is implemented below. While we need to define only the `post_randomize()` function to generate the parity bit, we also define the `pre_randomize()` function to show how it works:

```
1  class even_parity;
2     rand logic [4:0] data; // bit 4 is the parity bit
3
4     function void pre_randomize();
5       $display ("data before randomize: %5b", data);
6     endfunction
7
8     function void post_randomize();
9       data[4] = ^data[3:0];
10    endfunction
11 endclass
12
13 module top;
14
15    even_parity ep;
16
17    initial begin
18        ep = new();
19        for (int i=1; i<=4; i++) begin
20           ep.randomize();
21           $display ("data after  randomize: %5b", ep.data);
22           $display;
23        end // for
24     end // initial
25 endmodule
```

FIGURE 20-5. Using `pre_randomize()` and `post_randomize()`

- **Line 2**—We define a five-bit number.
- **Lines 4–6**—We use the `pre_randomize()` method to print the value of data before it's randomized.
- **Lines 8–10**—We use the `post_randomize()` method to do an XOR reduction of the bottom four bits (the ^ operator does this) and put the value into the top bit.
- **Line 15**—We define a handle of type `even_parity` called `ep`.
- **Lines 19**—We loop four times.
- **Line 20**—We randomize `ep`.
- **Line 21**—We display the results of randomization.

3. Well, actually we can, but that's in Chapter 21.

SystemVerilog calls the `pre_randomize()` method before it randomizes `data`. The method is trivial in this example, but in complex examples we might use it to do things such as populate dynamic arrays.

The `post_randomize()` method implements even parity. After `data` has been randomized, `post_randomize()` replaces its most-significant bit with the XOR reduction of the bottom four bits. The result is that we always have an even number of 1's.

Here's what happens when we run the simulation:

```
# vsim -c -sv_seed 5 top
# Loading sv_std.std
# Loading work.prepost_sv_unit
# Loading work.top
# data before randomize: xxxxx
# data after  randomize: 11000
#
# data before randomize: 11000
# data after  randomize: 01100
#
# data before randomize: 01100
# data after  randomize: 01001
#
# data before randomize: 01001
# data after  randomize: 10111
#
```

FIGURE 20-6. Creating Even Parity with `post_randomize()`

Notice that we are not creating a new `ep` each time through the loop, so the `pre_randomize()` method prints the value that was created the last time through the loop. There are always be an even number of bits set in these words, thanks to `post_randomize()`.

20.4 Seeding Randomization

Random-number generators don't generate random numbers. Instead, they generate a repeatable sequence of numbers that have no obvious pattern. These are essentially random numbers to the device under test.

It's a good thing that simulators don't generate truly random numbers, because then it would be impossible to debug our design. We'd see a bug, but we'd be unable to recreate it, because we couldn't reproduce the same sequence of data.

However, having a repeatable sequence of random numbers also has limitations. Once we've debugged everything from a given sequence, we will want to run the test again with a new sequence. This is where the *seed* comes in.

Random-number generators use the seed to calculate the first random number in a sequence of numbers. We can recreate a sequence of random numbers by reusing the same seed, and we can create a new list of random numbers by changing the seed. When we looked at `$random` we saw how to set the initial seed in Verilog '95. Now we need to do the same thing in SystemVerilog.

20.4.1 Setting the Seed at Runtime

Simulators allow you to use a command line option to set the seed when you start a simulation. For example, here is the Questa `-sv_seed` option:

```
% vsim -sv_seed <integer>
```

The simulator uses the base seed you've provided to initialize all random operations. If you call the simulator with the same seed number, you'll get the same results and you can create new tests by changing this seed.

20.4.2 Setting the Seed in an Object

Each object in SystemVerilog has its own, built-in, random-number generator. Therefore, the repeatable sequence of random numbers generated by each object is independent of the other objects in the simulation. This is important, as you don't want the random numbers being generated in an object to change just because you modified the behavior of an unrelated object.

Because each object has its own random-number generator, you can set the seed for each object separately. All SystemVerilog classes have a method that does this— the `srandom()` method.

SystemVerilog defines `srandom()` like this:

```
function void srandom (int seed);
```

You call `srandom()` as you would any other method. Let's define the seed in our `even_parity` example by using `srandom()`:

```
1  class even_parity;
2     rand logic [4:0] data; // bit 4 is the parity bit
3
4     function new (int seed);
5        $display;
6        $display ("setting seed to %0d in new()", seed);
7        $display;
8        srandom(seed);
9     endfunction
10
11    function void pre_randomize();
12      $display ("data before randomize: %5b", data);
13    endfunction
14
15    function void post_randomize();
16      data[4] = ^data[3:0];
17    endfunction
18 endclass
19
20 module top;
21
22    even_parity ep;
23
24    initial begin
25        ep = new(5);
26        for (int i=1; i<=4; i++) begin
27           ep.randomize();
28           $display ("data after  randomize: %5b", ep.data);
29           if (i == 1)
30           $display ("                       ^^^^^");
31
32           $display;
33        end // for
34        ep.srandom(5);
35        assert(ep.randomize());
36        $display;
37        $display ("resetting seed to 5");
38        $display ("data after  randomize: %5b", ep.data);
39        $display ("                       ^^^^^");
40        $display ("Note the same first number");
41
42     end // initial
43 endmodule
```

FIGURE 20-7. Setting Seeds in an Object

- **Lines 4–9**—We've modified the `even_parity` object so that its constructor takes a `seed` argument. We put `seed` into the random-number generator with `srandom()`.
- **Line 25**—We pass the constructor the number 5 to use for the call to `srandom()` in the constructor.
- **Line 30**—The `^^^^^` carets produce an attractive underline in the output.
- **Line 34**—We call `srandom()` from outside the object and restart the random-number generator with a seed of 5.

We can see the result below. We created `even_parity` with a seed of 5, so our first "random" number is `'b00011`. Then we manually reset the seed to 5 (in line 34 above) and we see that we get the number `'b00011` again:

```
# setting seed to 5 in new()
#
# data before randomize: xxxxx
# data after  randomize: 00011
#                        ^^^^^
#
# data before randomize: 00011
# data after  randomize: 01111
#
# data before randomize: 01111
# data after  randomize: 01001
#
# data before randomize: 01001
# data after  randomize: 01100
#
# data before randomize: 01100
#
# resetting seed to 5
# data after  randomize: 00011
#                        ^^^^^
# Note the same first number
```

FIGURE 20-8. Results of Controlling the Seed

Changing the seed between simulation runs moves your design in new directions and finds new bugs. It's so effective that I've heard of one story where a manager told his team to stop using new seeds—because they kept finding new bugs and he said they needed to ship.

Makes you wonder …

20.5 Turning Randomization On and Off

There may be times when you don't want to randomize all the variables in a class. For example, say you have a transaction that contains mode bits in a variable. Normally, you like to randomize the mode bits to throw a wide variety of stimulus at your DUT.

Now, suppose you realize that there is one test case where you want to set the mode bits and then randomize the other data based on the mode bit (you'll see how to do this in the coming chapters). In that case, you'd want to turn off the randomization for the mode bits but not for the other variables. You can turn off the randomization with the `rand_mode()` method.

SystemVerilog defines the `rand_mode()` method for every class that contains random variables and every random variable in a class. The method has two versions: a *task* version and a *function* version. Here are the definitions of each for `rand_mode()`:

```
task object[.random_variable].rand_mode(bit on_off);

function int object.random_variable.rand_mode();
```

The `rand_mode()` task has three major differences from the function:

- The task method requires an argument: 0 to turn randomization off, 1 to turn randomization on.
- The task method exists for both classes and random variables. The function method exists only in random variables.
- The function method has no argument. It returns a 1 if the random variable can be randomized or a 0 if it cannot.

The `rand_mode()` function exists only in variables that were declared with the `rand` or `randc` qualifier. If we try to call `rand_mode()` on an object or variable that doesn't have these qualifiers, we'll get a syntax error.

Here is an example of using the `rand_mode()` method both in its task and function modes. We are going to (1) randomize an object, (2) turn randomization off and try again, (3) turn randomization on for just two variables, and (4) try again.:

```
1  class my_rand_obj;
2     rand  logic [2:0] random;
3     randc bit   [2:0] cyclic;
4     randc enum {DOGS, CATS, TREES, FLOWERS} life;
5
6     function void pre_randomize();
7        $display (" ** randomizing **");
8     endfunction
9
10    function void print;
11       $display ("random: ",random);
12       $display ("cyclic: ",cyclic);
13       $display ("life: %s"  ,life);
14       $display;
15    endfunction
16
17
18 endclass
19
20
21 module top;
22
23    my_rand_obj r;
24
25   initial
26    begin
27       r = new();
28       assert(r.randomize());
29       r.print;
30
31       $display ("call: r.rand_mode(0);");
32       r.rand_mode(0);
33       $display ("r.rand_mode.random() -> %0b ",r.random.rand_mode());
34       $display ("r.rand_mode.cyclic() -> %0b ",r.cyclic.rand_mode());
35       $display ("r.rand_mode.life()   -> %0b,",r.life.rand_mode());
36       assert(r.randomize());
37       r.print();
38
39       $display ("call: r.random.rand_mode(1);");
40       $display ("call: r.life.rand_mode(1);");
41       r.random.rand_mode(1);
42       r.life.rand_mode(1);
43       $display ("r.rand_mode.random() -> %0b ",r.random.rand_mode());
44       $display ("r.rand_mode.cyclic() -> %0b ",r.cyclic.rand_mode());
45       $display ("r.rand_mode.life()   -> %0b,",r.life.rand_mode());
46       assert(r.randomize());
47       r.print();
48     end
49 endmodule
```

FIGURE 20-9. Using `rand_mode()`

- **Lines 10–15**—Create a `print()` method that writes the values in the object to the screen.
- **Lines 28–29**—Randomize the `my_rand_object` object `r` and print the result by using the `print()` method.
- **Line 32**—Use `rand_mode()` as a task to turn off all randomization in object `r`.

- **Lines 33–35**—Use `rand_mode()` here as a function to get the status of the three random variables. Naturally, their status is 0 because we just turned them off.
- **Lines 36–37**—Randomize and print again. Of course, nothing changes because we've turned off randomization for all the variables in the object.
- **Lines 41–42**—Use `rand_mode()` as a task to turn randomization back on for just two variables: `random` and `life`.
- **Lines 43–45**—Use `rand_mode()` as a function to get the status of the three random variables: `random` and `life` are now 1, `cyclic` remains 0.
- **Lines 46–47**—Randomize `r` and print it again. Only `random` and `life` change.

Here is the result of running this program:

```
#  ** randomizing **
# random: 1
# cyclic: 6
# life: CATS
#
# call: r.rand_mode(0);
# r.rand_mode.random() -> 0
# r.rand_mode.cyclic() -> 0
# r.rand_mode.life()   -> 0,
#  ** randomizing **
# random: 1
# cyclic: 6
# life: CATS
#
# call: r.random.rand_mode(1);
# call: r.life.rand_mode(1);
# r.rand_mode.random() -> 1
# r.rand_mode.cyclic() -> 0
# r.rand_mode.life()   -> 1,
#  ** randomizing **
# random: 4
# cyclic: 6
# life: FLOWERS
```

FIGURE 20-10. Turning Random Variables On and Off

We can see that the first randomization sets `random`, `cyclic`, and `life` to initial values. Then, when we turned all the variables off, we got no change on randomization. Finally, we turned two of the variables on and they changed when we randomized.[4]

4. Although they seem to stay the same if the are randomized back to the same values.

20.6 Summary

In this chapter, we examined the basics of randomizing an object. We randomize an object by using the `rand` and `randc` declaration keywords to identify random variables. Then we execute the `randomize()` method SystemVerilog has built into every object.

The `rand` and `randc` keywords control the kind of randomization we get. The `rand` keyword delivers random numbers spread uniformly over the range of possible values. The `randc` keyword cycles through all possible values for a variable and then starts over (as in a deck of cards.)

We change the behavior of `randomize()` with the `pre_randomize()` and `post_randomize()` methods. The `randomize()` method calls these before and after randomizing the variables.

SystemVerilog random numbers are not really random, as they depend on an initial seed. By changing the seed we can get different sequences of random numbers, and, as a result, find more bugs in our design. We learned how to set the seed with a command line option and with the `srandom()` method.

Finally, we can turn randomization off with the `rand_mode()` *task*, and check the randomization status of every variable with the `rand_mode()` *function.*

This may seem like a lot of randomization control, but it is still not enough control to create powerful test benches. To do that, we need to be able to constrain the numbers from our random-number generator with much finer granularity. We'll do that in the next chapter.

21 Constraining Random Variables

IN THIS CHAPTER

- Constraining Random Variables
- Using the `with` Statement
- Constraining with Expressions
- Constraining with Sets
- Constraining with Distributions

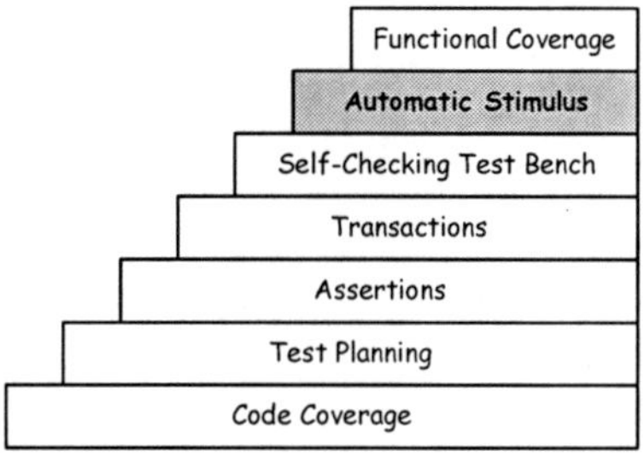

Writing tests is a time consuming task. In fact, the time needed to write tests is the primary argument for just taking an FPGA to the lab and powering it up. We counter this tendency by requiring that the engineers write enough tests to reach 100% code coverage (with a list of exemptions) before we go to the lab.

But, as we've seen in the past, writing directed tests focused on individual behaviors is an exercise in combinatorial explosion. Once we have a self-checking test bench, its easier to write longer and more complete tests, but this can still be time consuming. This is where randomized stimulus comes in. It allows the test bench itself to create the tests.

However, randomized stimulus has its own problem. It may be too random. For example, let's say that we want to test a system with several masters and several slaves, and we really want to focus on the case where two masters request the same operation from the same slave. If we've randomized the master and slave addresses, we'll rarely test the case we want. It would be much faster to constrain our data to force this to happen.

If we think about creating tests that use constrained random stimulus, we come up with a three-step approach:

1. Run a simulation with random stimulus and look at the coverage results. Change the seed and run again to see if this improves coverage. If we've still missed coverage, go to step 2.
2. Based on the missed coverage, create randomization constraints that force the simulation to cover the missed areas. Change the seed to see if this improves coverage. If we've still missed coverage, go to step 3.
3. Write directed tests to fill in the missing areas and achieve 100% coverage.

By following these three steps, we completely cover our design as quickly as possible, and we need to write only a few directed tests.

21.1 SystemVerilog Constraints

When we talk about constraints, we are talking about the ability to say something such as, "I want a random number between one and one hundred" or, "I want a random number that has even parity." Unlike Verilog, or VHDL, or even C++, SystemVerilog has constructs built into the language that allow us to constrain randomization in the ways I just mentioned.

SystemVerilog constrains random variables in two ways. Both involve creating constraints that limit the numbers from the random-number generators, but they work at different points in the process:

1. We can control `randomize()` when we call it by using the `with` keyword.
2. We can add constraints to the transactions themselves, so that they generate only the appropriate random data.

The industry argues constantly over which of these approaches is better. As usual, both have their strengths and weaknesses.

In our case, we'll use the `with` keyword to talk about the various ways one can constrain randomization. Then we'll add constraints to our transactions to create a test bench that writes its own tests.

21.2 The `with` Keyword

Until now, we've been calling `randomize()` and letting it create what it wants. While we've modified the random numbers by using `post_randomize()`, that's not as effective as constraining the random numbers as they are generated. Now we'll start constraining our numbers by using the `with` statement.

Here is how we call `randomize()`, constraining the results by means of the `with` statement, and testing the success by means of the `assert` statement:

```
assert(obj_handle.randomize() with {
    <a list of constraints>
});
```

We call `randomize()` within an `assert` statement and use the `with` keyword to designate constraints. The constraints allow us to say that the random numbers in `obj_handle` conforms to a set of rules that we designate.

We can make mistakes when we create the rules. For example, we might say that a data member in `obj_handle` called `A` needs to be greater than 10 and less than 3. This is, of course, impossible. In that case, `obj_handle.randomize()` returns 0 and `assert` creates an error message.

The `with` keyword assumes that we are working on variables within the `obj_handle` object. For example, let's say that `obj_handle` has a variable called `A` and we want to make sure that this variable is greater than 10. We'd write this:

```
assert(obj_handle.randomize() with {
    A > 10;
});
```

Notice that we did not type `obj_handle.A` in the constraint. The constraint statements assume we're talking about variables in `obj_handle`, since we called its `randomize()` method.

The line `A>10;` is a constraint, which SystemVerilog passes to a portion of the simulator called the *constraint solver*. The constraint solver reads all the constraint expressions and delivers a number that fulfills them. If the constraint solver can't fulfill the constraints, `randomize()` returns a `0`.

Now that we've seen our first constraint, it's time to look at constraints in more detail. We are going to start with the most straightforward.

21.2.1 Using the `with` Statement

When you call randomize in conjunction with a `with` statement, you need to supply a block of code called a *constraint block.* This block of code sits between two curly brackets (`{ }`) and contains expressions that control the random variables. There is no semicolon after the closing curly brace.

The expressions in the braces are called *constraint expressions.* Constraint expressions are boolean expressions that tell the constraint solver which numbers are allowed. The constraint solver succeeds only if it can find numbers that make all the expressions true. Here's an syntax sample for some object with two variables, `X` and `Y`:

```
assert(obj.randomize() with {
    <constraint expression 1 using X>;
    <constraint expression 2 using Y>;
    <constraint expression 3 using X and Y>;
});
```

This says, "Create random numbers in `obj.X` and `obj.Y` such that the three expressions are all true." If such a number cannot be created, then `obj.randomize()` returns a 0 and we get an error from the `assert` statement.

In the following sections we'll examine different types of constraint expressions.

21.3 Simple Expressions in Constraints

We're now ready to constrain our randomization and generate only the transactions we want. The techniques in this section allow us do things like force arithmetical relationships between our random variables, control parity on data, and limit the ranges of numbers.

We create constraints by verbally describing what we want and then figuring out how to create a set of expressions that capture our goal. The most basic constraint expressions are familiar to us from simple Verilog coding, for example:

Randomize two variables A and B such that
A is less than 'h4 and B is less than 'hA.

We have an object that contains A and B and we are going to randomize that object with a constraint block. Here is the code:

```
1  typedef enum {add_op, and_op, xor_op, mul_op,no_op,rst_op} operation_t;
2
3  class alu_operation;
4     rand logic[7:0] A,B;
5     rand operation_t op;
6  endclass
7
8
9  module top;
10    alu_operation req;
11
12 initial begin
13      $display ("---- A < 'h4 , B < 'hA ----");
14      req = new();
15      for (int i = 1; i<=5; i++) begin
16          assert(req.randomize() with {
17                A < 'h4;
18                B < 'hA;
19            });
20          $display("A: %3h  B: %3h   SUM: %2h", req.A, req.B,req.A+req.B);
21      end
```

FIGURE 21-1. Constraining with Simple Expressions

- **Lines 1**—Declare an enumerated type called `operation_t`. It has six members. We use this in a later example.
- **Lines 3–6**—This is our class definition for `alu_operation`. It says that `A`, `B` and `op` can be randomized.
- **Line 10**—Declare `req` to be an `alu_operation` handle.
- **Line 14**—Create a new `alu_operation` object and put the handle into `req`.
- **Line 15**—Randomize `req` five times.
- **Line 16**—Call the `randomize()` method from within an `assert` statement by using a `with` keyword. The `with` keyword uses curly brackets to contain the constraint expressions.
- **Lines 17–18**—These are our constraints. We want the `A` variable to be less than `'h4`, and the `B` variable to be less than `'hA`.
- **Line 20**—Display the results.

When we run this code, it creates an object called `req` and randomizes it five times so that `A` is always less than `'h4` and `B` is always less than `'hA`.

Here are the results:

```
# ---- A < 'h4 , B < 'hA ----
# A:  02  B:  00   SUM: 02
# A:  01  B:  09   SUM: 0a
# A:  01  B:  08   SUM: 09
# A:  02  B:  05   SUM: 07
# A:  03  B:  08   SUM: 0b
```

FIGURE 21-2. Randomizing A and B with Constraints

The key to the constraint expressions is that all the expressions need to be true for randomize to return a number. The SystemVerilog *Language Reference Manual* (LRM) says that SystemVerilog simulators must always return a number that matches the constraints if that is possible.

Since constraint expressions are boolean expressions, we could have written this expression on one line:

```
(A < 'h4 && B < 'h10)
```

However, I don't recommend combining expressions this way. Constraint solvers work by analyzing the constraint lines and then removing numbers from the possible set of results. They work much faster if they can work on one simple constraint line at a time. So it is better, from a performance perspective, to break constraints up into as many simple lines as possible.

The expressions in Figure 21-1 are simple boolean declarations, but we are not limited to these. We can use more-complicated expressions in constraints. For example, we can create the following:

Randomize A and B such that their sum equals 8'hFF.[1]

This is a corner case that we may want to be sure to check, the maximum output of an adder without a carry. Rather than write a directed test to create this kind of output, we can just ask SystemVerilog to create the correct stimulus.

1. The number 8 denotes the width of the constant in Verilog and SystemVerilog.

SystemVerilog can use any operation when calculating a set of random variables. So you could also ask for outputs that mask certain bits, or force the inputs to the adder to create a number greater than `9'h100` to test the carry out.

Here's how to make SystemVerilog create inputs that sum to `8'hFF`:

```
      $display ("---- A + B == 8'hFF ----");
      for (int i = 1; i<=5; i++) begin

        assert(req.randomize() with {
              A+B == 8'hFF;
            });
          $display("A: %3h  B: %3h    SUM: %2h", req.A, req.B,req.A+req.B);
      end
```

FIGURE 21-3. Randomizing to a Sum

SystemVerilog will find solutions so that the expression is true in all cases. Here is how it responded to this constraint:

```
# ---- A + B == 8'hFF ----
# A:  a4  B:  5b   SUM: ff
# A:  17  B:  e8   SUM: ff
# A:  f8  B:  07   SUM: ff
# A:  4f  B:  b0   SUM: ff
# A:  5d  B:  a2   SUM: ff
```

FIGURE 21-4. `A` and `B` Randomized to Sum to `8'hFF`

SystemVerilog had no problem fulfilling this request.

We can use any combination of expressions in this way and SystemVerilog simulators finds suitable numbers if possible. But, there are limits to what you can do with logical and arithmetic expressions so we'll move on to new ways of defining constraints.

21.4 Constraining Random Variables with Sets

What would we do if we wanted our random variable to be one of five different numbers? With expressions we'd need to create a long logical expression with

ANDs and ORs to make that happen. It would be even worse if we wanted to have our numbers fall within a range.

We can avoid all that unpleasantness by using the `inside` operator in our constraint. It looks like this:

```
<expression> inside {<set of numbers>};
```

This constraint says that the result of the expression before the keyword `inside` needs to fall within the set of numbers that follows. Here's an example of using `inside`:

Create random variables A and B such that A+B is either equal to 8'h10 or 8'hFF.

Using the same `req` object we used in the last example, we can implement this constraint with the following code:

```
26        $display ("---- A + B inside {8'h10,8'hFF} ----");
27        req = new();
28        for (i = 1; i<=5; i++) begin
29          assert( req.randomize()  with {
30              (A + B) inside {8'h10,8'hFF};
31            });
32          $display ("A: %2h  B:%2h   A+B:%3h", req.A,req.B,req.A+req.B);
33        end // for (i=1; i<=5; i++);
```

FIGURE 21-5. Using the `inside` Operator

The `inside` operator (on line 30) has some tricky syntax. The problem is that the `inside` operator uses the same curly braces, `'{}'`, for set membership that the `with` operator uses to encapsulate a set of constraints[2]. This leads to confusion when it comes to the use of semicolons.

The `inside` operator curly braces need a semicolon at the end of every line, because all constraint expressions end with a semicolon. The curly braces connected to the `with` statement don't use a semicolon, so sometimes it is hard to remember where to put the semicolon. Line 30 could have been written this way:

```
(A+B) inside {8'h10,8'hFF};}
```

2. Not to mention that the rest of Verilog uses it for concatenation and replication. It's a very popular character.

The syntax results in a strange "right-brace sandwich" around the semicolon.[3] The result is as we expect. We get combinations that sum to both numbers:

```
# ---- A + B inside {8'h10,8'hFF} ----
# A: c8  B:37   A+B: ff
# A: 89  B:76   A+B: ff
# A: 7b  B:95   A+B: 10
# A: bb  B:55   A+B: 10
# A: 5e  B:a1   A+B: ff
```

FIGURE 21-6. Result of Using the `inside` Operator

21.4.1 Ranges in Sets

In our last example, we used two individual numbers as a target set for the constraint. We can do more than use individual numbers to define sets. We can also use ranges.

Let's say that we have an router with a corner case that worries us. If we want to force the test to exercise that corner case, we can make a complex constraint like this:

Create random variables A and B such that
A is between 0 and 'h5 or equal to 'hFF and
B is between 'h0A and 'h0F or between 'h01 and 'h03 or equal to 'hFE.

We implement ranges like "between `0` and `'h5`" with square brackets. The phrase "between 0 and `'h5`" becomes `[0:'h5]`. We can put multiple ranges in the same constraint to implement things like the `B` constraint above, where `B` must be within one of two ranges.

3. This language idiosyncrasy reminds me of the ladder on the left-field wall at Fenway Park in Boston. It's one of those charming imperfections, left over from a previous time, that make life interesting.

The square brackets let us implement these range constraints like this:

```
16  $display ("---- A inside {[0:8'h5], 8'hFF} ----");
17  $display ("---- B inside {[8'h0A:8'h0F], [8'h01:8'h03], 8'hFE} ----");
18  for (i = 1; i<=5; i++) begin
19    assert(req.randomize() with {
20        A inside {[0:8'h5],8'hFF};
21        B inside {[8'h0A:8'h0F], [8'h01:8'h03], 8'hFE};
22      });
23    $display ("A: %2h  B:%2H", req.A,req.B);
```

FIGURE 21-7. Ranges in Set Constraints

- **Line 20**—The square brackets here create a range for `A`.
- **Line 21**—The two sets of square brackets create two ranges.

Notice that we can mix ranges and single numbers to create a constraint. For example `A` can be in the range `0-'h5` or equal to `'hFF`. When we run this example we get the following results:

```
# ---- A inside {[0:8'h5], 8'hFF} ----
# ---- B inside {[8'h0A:8'h0F], [8'h01:8'h03], 8'hFE} ----
# A: ff  B:03
# A: 02  B:0b
# A: 00  B:0a
# A: 02  B:03
# A: 02  B:0f
```

FIGURE 21-8. Results of Sets

In these examples I created ranges that did not overlap, but this doesn't need to be the case. You can say that `A` needs to be inside of the range `{[1:10],[5:15]}`. SystemVerilog creates a union of these ranges.

Ranges can run between any two legal numbers for the variable in question, but you can also designate a range to run up to the maximum value for a variable. For example, our 8-bit `A` variable has a maximum value of `8'hFF`. However, we might want to create a range that works for any width. In that case, we want to constrain `A` from some number to the maximum possible. We do that with the maximum operator: `$`.

```
26        $display ("---- A inside {[8'hFB:$]} ----");
27        for (i = 1; i<=5; i++) begin
28          assert(req.randomize() with {
29              A inside {['hFB:$]};
30          });
31          $display ("A: %2h", req.A);
32        end // for (i=1; i<=5; i++);
```

FIGURE 21-9. Using the Maximum Operator: `$`

- **Line 29**—The maximum operator designates the numbers `'hFB` to `'hFF` because `A` is eight bits wide.

Here's what we see:

```
# ---- A inside {[8'hFB:$]} ----
# A: fc
# A: fe
# A: fe
# A: fe
# A: ff
```

FIGURE 21-10. Results of Maximum Operator in Range

We can also use ranges with enumerations. SystemVerilog defines an order for enumerated values based on the enumeration definition. For example, here is the enumeration that defines the operations for our TinyALU:

```
1  typedef enum {add_op, and_op, xor_op, mul_op,no_op,rst_op} operation_t;
2
3  class alu_operation;
4     rand logic[7:0] A,B;
5     randc operation_t op;
6  endclass
```

FIGURE 21-11. Enumeration Definition for `operation_t`

- **Line 1**—Define an enumerated type with the TinyALU Operations.
- **Line 5**—Use the enumerated type to help define the transaction called `alu_operation`.

SystemVerilog enumerations assign integers to the values in the enumeration list. Consequently, in this example `add_op` has the integer value `0` and the integers increment up to `rst_op` which has the integer value `5`.

This allows us to create a range with a randomized enumeration. For example we can say that we want only operations that range from `add_op` (0) to `xor_op` (2):

```
58        $display ("---- op  inside [add_op : xor_op] ----");
59        for (i = 1; i<=6; i++) begin
60           assert(req.randomize() with  {
61                op inside {[add_op : xor_op]};
62             });
63             $display ("OP:%s", req.op);
64         end
```

FIGURE 21-12. Using Enumerated Types in a Range

- **Line 46**—Force `op` to be in the range from `add_op` to `xor_op`.

Also, notice that Figure 21-11 on page 259 declares `op` to be a cyclic random variable by using the `randc` keyword. This means that we should exhaust all the possibilities between `add_op` and `xor_op` before repeating ourselves.

The combination of the range constraint and `randc` does just what we expect. We see all three possible operations before the randomization repeats itself:

```
# ---- op  inside [add_op : xor_op] ----
# OP:add_op
# OP:xor_op
# OP:and_op
# OP:and_op
# OP:add_op
# OP:xor_op
```

FIGURE 21-13. Cyclically Random Enumerated Types

This is a great example of `randc` in action. The operations cycle through the operations `add_op`, `xor_op`, then `and_op`. Now, with all three types of operations exhausted, the simulation randomly picks a new starting point `and_op` and randomizes from there.

You can combine the `inside` keyword with other expressions. For example, you could create addresses in the range from `16'h10` to `16'h20` and force the least-significant bit to be `0` by combining an expression and a range on two lines:

```
66        $display ("---- A  inside [`h10 : `h20] ----");
67        $display ("----            A[0] == 0;        ----");
68        for (i = 1; i<=6; i++) begin
69           assert(req.randomize() with  {
70                A inside {['h10 : 'h20]};
71                A[0] == 0;
72           });
73           $display ("A: %2h", req.A);
74        end
```

FIGURE 21-14. Combining Ranges with other Expressions

- **Line 70**—Force A to be a number between `'h10` and `'h20`.
- **Line 71**—Force A to be an even number.

This gives us the following results:

```
# ---- A  inside [`h10 : `h20] ----
# ----            A[0] == 0;       ----
# A: 20
# A: 1e
# A: 12
# A: 14
# A: 12
# A: 1a
```

FIGURE 21-15. Generating Even Numbers Beween `` `h10 `` and `` `h20 ``

The `inside` operator limits the range of numbers we can generate, but all the numbers have the same probability of being generated. We can control the distribution with distribution constraints.

21.5 Distribution Constraints

When we limit the possible values of a random variable by using expressions or ranges, we create a set of possible values for the variable. For example, when we set the 8-bit value `A` to be in the range of `8'hFB` to `8'hFF`, we were limiting `A` to a set of five possible values.

By default, SystemVerilog gives each value in the set an equal chance of being chosen. In this case, the five possible values of A each have a 20% (1/5) chance of being chosen when we called `randomize()`.

However, what if we don't want to have an equal chance for all values. Consider the enumerated type `operation_t` in our TinyALU example. We can see in Figure 21-11 on page 259 that there are six possible operations, one of which is `rst_op`. Do we really want to write a test bench that has a one-in-six chance of resetting the design each time we generate a transaction? Probably not. However, we do want to see an occasional reset.

SystemVerilog lets us control the distribution of values with the `dist` keyword. The `dist` keyword works like the `inside` keyword, but it designates the odds that any of the elements of the set will be chosen. It works with the := operator and :/ operator to allow us to define probabilities. We use these operators to map probabilities to sets in SystemVerilog.

We'll examine distribution with a completely different randomization problem: sports (or, for those who speak English instead of American: sport)

Let's say that I want to choose randomly between four sports:

```
baseball, football[4], hockey, basketball
```

If I want to choose among them with even odds I need to give each one a probability such that all probabilities add up to 100%:

```
baseball(25%), football(25%), hockey(25%), basketball(25%)
```

Now, say that I really like baseball, so I want to give it a 50% probability. I need to divide the remaining probability among the other sports:

```
baseball(50%), football(16.67%),
hockey(16.67%), basketball(16.66%)
```

I was forced to do a little bit of calculation to create probabilities that add up to 100%.

Now we add lacrosse to the mix.

4. In the spirit of internationalization, the reader can choose whether this means American football.

```
baseball(50%), football(12.5%), hockey(12.5%),
      basketball(12.5%), lacrosse(12.5%)
```

We see here that adding a new value has forced me to recalculate all the other values. Then it gets more complicated if I don't like lacrosse and want to set it to 5%:

```
baseball(50%), football(15%), hockey(15%),
      basketball(15%), lacrosse(5%)
```

This kind of recalculation is tedious and error prone. We would prefer it if SystemVerilog could just do these calculations for us and, fortunately, it can. SystemVerilog uses a system of weights to allow us to designate the probabilities without our needing to do the calculations; it uses the := and :/ operators to connect those weights to values. Let's revisit these examples, this time using SystemVerilog operations. We start with an enumerated type:

```
typedef enum {baseball, football,
  hockey, basketball} sports_t;
```

Assuming we have a random variable called `sport;` we want to control the probabilities of any of the sports being selected in a `randomize()` call. To do this, we add constraints by using a `with` statement. We just focus on the constraints in this section, because we know the rest of the syntax.

In our first case, we want to choose between the four sports with an equal chance of selecting each. Assume that we have an object called `play` that contains a variable called `sport` of type `sport_t`, we specify the equal distribution like this:

```
16          play.randomize() with {
17           sport dist {baseball:= 1, football:=1, hockey:=1, basketball:=1};
18        });
```

FIGURE 21-16. Even Distribution Among Sports

Notice the `dist` constraint on line 17. The syntax looks a lot like the syntax associated with the `inside` operator, but now we have weights assigned to each option in the set. The := operator gives each choice here a weight of 1.

In this example there are four weights of 1, for a total weight of 4, so each value has a 1/4 chance of being chosen. This is exactly the same as if we had not constrained `sport` at all.

Let's change the weights to accommodate my love of baseball. I want `baseball` to have a 50% chance of being chosen. This doubles its chances, so one might think that this would be the way to do this:

```
sport dist {baseball:=2, football:=1, hockey:=1,
basketball:=1};
```

One would be wrong. Because now we have a total weight of 2+1+1+1 for a total of 5. The chance of baseball being chosen is now 2/5 or 40%—not 50%. To get 50% we need to do this:

```
        play.randomize() with {
         sport dist {baseball:= 3, football:=1, hockey:=1, basketball:=1};
      });
```

FIGURE 21-17. Doubling the Chances for Baseball

Now the total weight is 3+1+1+1 for a total of 6, and `baseball` has three chances in six of being chosen.

It's tedious to write := over and over if several values have the same weight. There is a better way. We can create a range and use the := operator to assign a weight to the range.

We now want baseball to have a weight of 3 and all the others to have a weight of 1, so we create a range of the other values and assign them all a weight of 1:

```
        play.randomize() with {
            sport dist {baseball:= 3, [football:basketball]:=1};
      });
```

FIGURE 21-18. Using a Range with `dist`

However, we still have a problem. We are still going to get our randomization screwed up if we modify our enumerated type. For example, let's add `lacrosse` to the list of sports. We'll put it at the end of the enumeration called `sports_t` and modify our constraint. Then we have this:

```
        play.randomize() with {
            sport dist {baseball:= 3, [football:lacrosse]:=1};
    });
```

FIGURE 21-19. Adding `lacrosse` to the Distribution

Now we've lost the 50% for `baseball` again. There are four sports in the range, so we have a total weight of 3+1+1+1+1 for a total of 7 and the probability of `baseball` being chosen is 3/7. So that's wrong. To get baseball back to a 50% chance, we would need to change its weight to 4. What a pain.

SystemVerilog has an operator that can help solve this problem—the :/ operator. The slash (/) in the operator reminds us that this operator divides the weight across all the members of the range.

Whereas the := operator says that every value in the range gets the weight after the operator, the :/ operator says that the *entire range* gets the weight. It divides the weight by the number of items in the range and gives each item in the range the new, divided, rate. Here's how :/ looks in action:

```
        play.randomize() with {
            sport dist {baseball:= 1, [football:lacrosse]:/1};
    });
```

FIGURE 21-20. Using the :/ Operator

This assigns a weight of 1 to the entire range of `football` to `lacrosse`. It says, "We have 1 chance of getting `baseball` and one chance getting a value from the range set." So `baseball` has a 50% chance regardless of the number of elements in the range.

Here are the results of a SystemVerilog program that randomly choose a game 10,000 times with different constraints, and shows how many times we chose `baseball`:

```
# ---- {baseball:= 1, football:=1, hockey:=1, basketball:=1} ----
# We played baseball 2490 of 10000 times (24.90 %)
# ---- {baseball:= 3, football:=1, hockey:=1, basketball:=1} ----
# We played baseball 4981 of 10000 times (49.81 %)
# ---- {baseball:= 3, [football:basketball]:=1} ----
# We played baseball 5007 of 10000 times (50.07 %)
# ---- {baseball:= 3, [football:lacrosse]:=1} ----
# We played baseball 4383 of 10000 times (43.83 %)
# ---- {baseball:= 1, [football:lacrosse]:/1} ----
# We played baseball 4986 of 10000 times (49.86 %)
```

FIGURE 21-21. Results of Distribution

We got close to our expected distribution in all these runs of 10,000 randomizations.

21.6 Summary

This was a big chapter, but it delivered the most powerful tool in our toolbox for automatic test generation. We learned how to constrain random variables by using expressions. These can be as simple as "A must be less than 10," or they can be complicated combinations of logical expressions and mathematical results.

We learned about the `with` statement and how we could use it to attach constraints to a call to `randomize()`. We saw how to use the `assert` statement to catch cases where our constraints made it impossible to deliver a random number.

We learned how to use the `inside` operator to describe sets of numbers in the constraints. We saw that we could combine the inside operator with expressions so that we can say things such as, "A + B must be inside the set of numbers from `'h0FE` to `h'105`".

Finally, we learned how to use distributions to change the odds that the constraint solver will choose a value or set of values. We saw that we could control weights of individual values or distribute weights across sets of values.

Simple expressions are powerful constraint tools, and they can control mathematical or logical relationships among variables. However, there are cases where we want to apply different constraints to one variable depending on the value of another variable.

We do that with conditional constraints, the subject of the next chapter.

22 Conditional Randomization

IN THIS CHAPTER:

- *The Implication Operator:* `->`
- *Controlling Constraints with* `if...else`

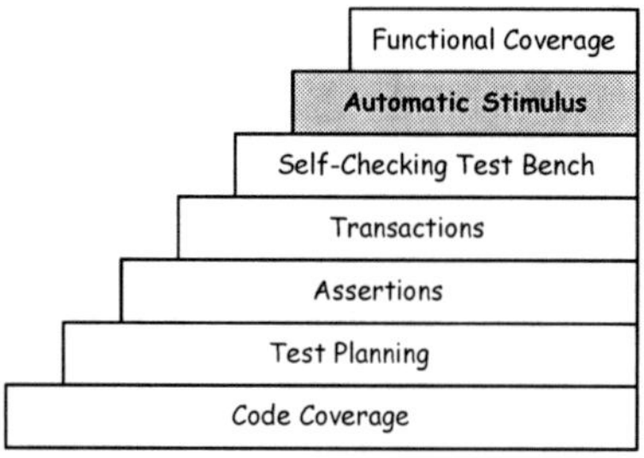

SystemVerilog has a powerful constraint description system. We saw the system's foundation in the previous chapter when we examined expression-based constraints. Now we'll continue our walk through the SystemVerilog constraint system in this chapter and the next one.

In this chapter we talk about *conditional constraints*. There are many times when we need to control which constraints get used, depending on the value of data in a transaction. These are conditional constraints.

For example, we may have a design that reads from a memory range that is partly read-only. Because it would make no sense to create transactions that tried to write to a read-only memory range, we would need to force SystemVerilog to choose read operations only when the address is in the ROM range.

We may also have cases where one set of constraints makes sense only for a certain value. Let's say we have two bus interfaces: one of them supports burst mode and the other does not. We want to make sure that our random transactions call for burst mode only if we are accessing the interface that supports it.

We solve these problems by using conditional constraints. Conditional constraints implement one of two statements:

1. If condition A is true, then condition B must be true.
2. Use this constraint only when condition A is true.

We implement these two statements with an operator and a statement. The implication operator (->) handles the case where B must be true if A is true, and an `if` statement handles the case where we want to use a constraint conditionally.

22.1 The Implication Operator: ->

If it is raining, we don't play baseball.

This simple statement is an *implication*. The truth of the first statement implies the truth of the second. We are saying that the state of the weather (raining or not) controls whether we play baseball (or not).

SystemVerilog captures constraint implications with the -> operator (a dash and a greater-than sign.) The arrow of the operator captures the fact that implications are one way only. We can say that if it is raining we won't play baseball, but we cannot say that if we are not playing baseball it must be raining.

When we randomize variables, the simulator makes sure that the implication statement is true for all combinations of random values. Assuming that we have an enumerated random variable called `weather`, we capture our constraint like this:

```
18          play.randomize() with {
19            (weather == rain) -> (sport != baseball);
20          });
```

FIGURE 22-1. Rain Implies No Baseball

The constraint on line 19 says that if the expression on the left is nonzero, then the expression on the right must be nonzero. If there is rain, we don't play baseball. As in real life, we could play other sports when there is no rain, but we know that there's no baseball game that day.

22.1.1 The Implication Operator's Surprising Behavior

SystemVerilog simulators have a big job when it comes to constraint solving. They need to give us legal combinations out of the billions available. Their approach to solving this problem for implication constraints can lead to surprising behavior.

To examine that behavior, let's consider a simple constraint between a one-bit variable and a four-bit variable. Given that we have a one-bit variable called `flag` of type `bit` and a four-bit variable called `value` of type `logic`, we can make the following rule:

If `flag` *is one, then* `value` *is zero.*

This is an implication. The four-bit variable `value` can be randomized to any value from `4'h0` to `4'hF` unless `flag` is set to 1. In that case, the constraint solver forces `value` to `4'b0`.

Here is the code that creates and tests this implication:

```
1  class implication;
2     rand bit flag;
3     rand logic[3:0] value;
4  endclass
5
6  module top;
7     implication imp;
8     int flagcnt = 0;
9     initial begin
10      imp = new();
11      for (int i = 1; i<=10; i++) begin
12       assert (
13           imp.randomize() with {
14             flag -> (value == 0);
15           });
16       $display ("flag %b, value %h", imp.flag, imp.value);
17       flagcnt = flagcnt + imp.flag;
18      end
19      $display ("flagcnt: %0d",flagcnt);
20    end
21 endmodule // top
```

FIGURE 22-2. Using the -> Operator

- **Lines 1–4**—Define a class with two random variables: `flag` and `value`.
- **Line 7**—Declare a variable called `imp` to hold the implication class.
- **Line 10**—Create a new object and it into `imp`.
- **Lines 11**—Randomize `imp` ten times.

- **Line 14**—Use the implication operator to force `value` to `4'b0` when `flag` is set.
- **Line 16**—Display the value of `flag` and `value`.
- **Line 17**—Count the number of times `flag` was set.
- **Line 19**—Print the number of times `flag` was set.

This code is completely correct and legal, but the results may not be what you expected. Remember that the constraint guarantees that all random values make the implication statement true. However, it implements this guarantee in a surprising way that creates a distribution that is different from what you might expect.

Here is the (probably unexpected) result. Check out `flagcnt` and compare that count to the count you'd expect from flipping a coin:

```
# flag 0, value e
# flag 0, value 7
# flag 0, value 9
# flag 0, value 8
# flag 0, value 6
# flag 1, value 0
# flag 0, value b
# flag 0, value e
# flag 0, value 3
# flag 0, value b
# flagcnt: 1
```

FIGURE 22-3. Result of `flag -> (value == 0)`**. Hmmm** ...

All the results make the implication true, but there is something odd about `flag`'s distribution. We defined `flag` to be a `bit`, and we didn't constrain it in any way, so you'd think that randomizing `flag` would be like flipping a coin. We expect a 50% chance of getting 1.

In this example, we randomized a single bit ten times and got 1 once. What are the odds of that? We can use a binomial distribution calculator[1] to see that the odds of getting heads once in ten tries is about 1%. Later, I rewrote this code and randomized the implication a thousand times, and `flag` was set 63 times. The odds of flipping a coin a thousand times and getting only 63 heads is 6.8E-20. Something is amiss.

1. There is a link to one on www.fpgasimulation.com. You can also use it to find the odds that a baseball player that normally hits .300 would only be hitting .220 in his first 50 at bats. Is he injured or just unlucky? But I digress...

The problem is that we are thinking like humans and not like a constraint solver. Humans think about the problem this way:

Randomize the variables and throw out bad results.

However, constraint solvers can't work that way. The number of potential misses would be huge and the performance hit would be unacceptable. Instead, constraint solvers figure out all the legal combinations and then choose randomly from among them.

When the constraint solver is presented with an implication constraint like this,

```
flag -> (value == 0)
```

it recognizes that there are 32 possible ways to combine `flag` and `value`, because if you concatenate them you get a five-bit number.

The constraint solver creates a table of all 32 values (see Figure 22-4 on page 274) and removes the ones that don't fit the constraint. Then it randomly chooses from the remaining options. This is much more efficient than choosing random values and throwing most of them away.

Earlier, I said that you were better off giving a constraint solver a set of small constraints, rather than combining all constraints into one big constraint. This remove-and-then-choose behavior is the reason why we use small constraints.

If you give the constraint solver a set of small equations, it can easily evaluate each one and use it to shrink the pool of valid results. On the other hand, if you give the constraint solver a large, complex constraint, your simulation slows down because the constraint solver must understand the implications of the whole equation at once—rather than understanding each small equation and applying those equations incrementally.

In our case, it's easy to watch the constraint solver in action, because we have only 32 values to track.

Here is a table of all the possible combinations between `flag` and `value`. The values removed by the constraint solver are gray:

flag	value
0	0
0	1
0	2
0	3
0	4
0	5
0	6
0	7
0	8
0	9
0	A
0	B
0	C
0	D
0	E
0	F
1	0
1	1
1	2
1	3
1	4
1	5
1	6
1	7
1	8
1	9
1	A
1	B
1	C
1	D
1	E
1	F

FIGURE 22-4. Constrained Values

There are 17 legal combinations from a possible universe of 32. You can use all the combinations where `flag` is 0 and the one combination where `flag` is 1 and `value` is `4'b0`.

This means that we don't have a 50% chance of seeing `flag` set to 1. Instead, we have 1/17 or ~6% chance. It's no wonder we only saw 1 once in ten tries. If the odds are 1/17, our binomial calculator tells us that the odds of getting exactly a single 1 value for `flag` out of ten tries is 34%. The odds of one or zero times is 89%.

This understanding of how the constraint solver works makes it easier for us to write efficient constraints. Once we understand that SystemVerilog looks at the entire possible solution space and then carves pieces off, we can write constraints that make this job easier and that get our expected results.

If we want the "human" behavior, we can force SystemVerilog to give us our 50% chance at `flag` being set with the `solve...before` constraint.

22.1.2 The `solve...before` Constraint

As it implies, the `solve...before` constraint forces SystemVerilog to solve one random variable before the others. This constraint goes into the constraint block and looks like this:

```
solve <variable list> before <variable list>;
```

This constraint can appear anywhere in the constraint block and it still determines the order in which the constraint solver randomizes the variables. There are a few rules for using `solve...before`:

- It works only on variables declared with the `rand` qualifier and not with `randc`. Variables declared with `randc` are always solved before the others.
- The variables cannot be floating-point types.
- It is not fair to torment the simulator by saying "`solve a before b`" and then "`solve b before a.`" Unlike the computers on Star Trek, the simulator is not destroyed by the paradox. It just gives you an error.
- Any variables that aren't ordered are solved together.

The `solve...before` constraint says that we randomize `flag` before we randomize `value`. This means that `flag` gets a 50% chance of being 1, and then value will respond to the state of `flag`.

This creates a two-step effect in our constraints. We still have 32 possible results, but if we solve `flag` first, we'll cut that state space in half before we need to figure out `value`.

Here's the incremental approach in action. We start out with all 32 possible values, and then we pick a number for `flag`. This means that we randomly choose between the top list in which `flag` is 0 and the bottom list where `flag` is 1.

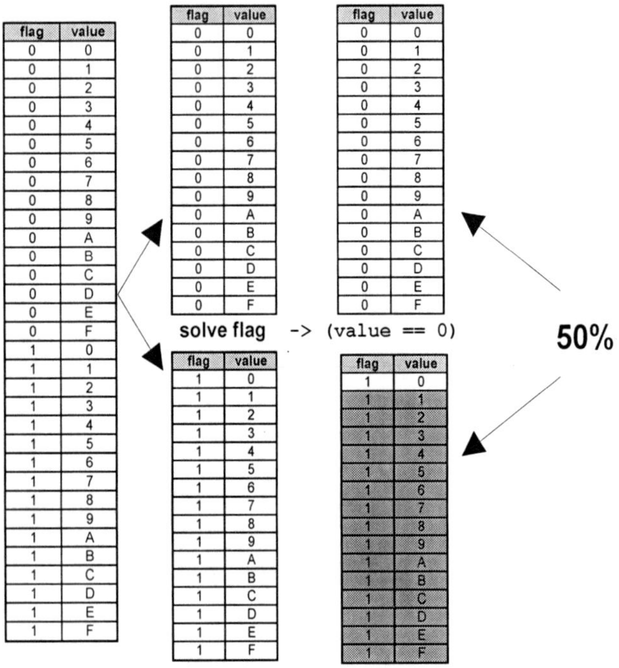

FIGURE 22-5. Constraints with Ordered Solving

Once we've chosen either the top list or the bottom list, we apply the rest of the constraint. If we choose the list with `flag` set to 0, the constraint solver can choose from among 16 values. Conversely, if we choose the list with `flag` set to 1, the constraint solver has only one choice.

This gives us a slightly greater than 50% chance that `value` will be `4'h0`. There is a 50% chance of `flag` being either 0 or 1. Then there is a 1/16 chance that `value` will be `4'h0` if `flag` is zero and a 100% chance if `flag` is 1. Therefore, the total chance that `value` will be 0 is calculated as follows:

$$(.5) \times (1) + (.5) * (1/16) = 0.53125$$

We implement this 50/50 chance for flag by including the `solve...before` constraint in our constraint block, like this:

```
13          imp.randomize() with {
14            flag -> (value == 0);
15            solve flag before value;
16          });
```

FIGURE 22-6. Using the `solve...before` Constraint

Line 15 causes `flag` to be solved first and gives us the following result:

```
# flag 1, value 0
# flag 0, value e
# flag 1, value 0
# flag 0, value f
# flag 0, value 3
# flag 0, value 4
# flag 1, value 0
# flag 0, value 6
# flag 1, value 0
# flag 1, value 0
# flagcnt:           5
```

FIGURE 22-7. Result of Solving `flag` First

Now `flag` is set to 1 five times. This is what we would expect.

22.2 Controlling Constraints with `if...else`

The `if...else` constraint allows you to turn constraints on and off based upon the status of a variable in the design. It is related to the implication constraint in that it can control the value of one variable depending upon the value of another. It is different in that it can completely turn constraints on and off depending on conditions.

The `if...else` construct makes it easier to capture constraints that may become too complicated to describe with implication constraints. Also, by completely ignoring unneeded constraints, you make it easier for the constraint solver to find solutions and thus your simulation runs faster.

The `if...else` constraint looks like this:

```
if (<logical expression>)
   <constraint set>;
else
   <constraint set>;
```

The constraint sets can either be one constraint or multiple constraints within curly braces—{}.

As an example, let's look at a transaction that represents a date. The object contains a month, a day, and a year. If we want to randomize this object, we need to take three things into account:

- Some months have 31 days.
- Some months have 30 days.
- February has 28 days some years and 29 days other years.

It would be possible to capture all these constraints in implications, but it would be complicated. It would be especially complicated to capture the leap year equations in implications. Instead, we can handle this more simply with an `if...else`. The `if...else` construct also allows us to nest conditions in a way that would be difficult to do with implications.

We define our transaction like this:

```
typedef enum {jan, feb, mar, apr, may, jun, jul, aug, sep, oct, nov, dec}
  month_t;

class date;
   rand integer year;
   rand month_t month;
   rand integer day;
endclass // implication_test
```

FIGURE 22-8. Definition of the `date` Transaction

Given the transaction we have above, we can create legal, random dates between 2001 and 2050, like this:

```
16          d.randomize() with {
17            solve year,month before day;
18            day >0;
19            year > 2000; // avoid the div by 400 non-leap year
20            year <= 2050;
21            if (month inside {jan, mar, may, jul, aug,oct,dec})
22               day <= 31;
23            if (month == feb)
24                if (year % 4 == 0)
25                   day <= 29;
26                else
27                   day <= 28;
28            if (month inside {sep, apr, jun, nov})
29               day <= 30;
30            });
```

FIGURE 22-9. Generating Dates with `if...else`

- **Line 16**—We are randomizing a date object called `d`.
- **Lines 17**—We want to choose `month` and `year` before we calculate `day`. We don't need this line, but it makes life easier for the constraint solver. With two of the variables set, the solver can focus on a smaller number of constraints. It's a good idea to use as much human intelligence as possible when creating constraints.
- **Line 18**—There cannot be a zero or negative `day`.
- **Line 19**—We want dates only after 2000.
- **Line 20**—We want dates only before 2051. Notice that I broke the possible range of the year into two constraints. This makes life easier for the constraint solver.
- **Lines 21–22**—We use `month` to allow `day` to be 31 for the long months.
- **Lines 23–27**—February is a lot of work and it forces us to nest conditional constraints. If we have a leap year we set the maximum day to 29, otherwise we set it to 28.

  Notice that the `else` statement always refers to the most recent `if` statement. This avoids confusion when we have nested `if...else` constructs.
- **Lines 28–29**—"*30 days hath September...*"

This example shows the advantage of the `if...then` constraint. Once `month` has been chosen, only one of the three `day` constraints is in effect. This makes it easier for the constraint solver—and a human reader.

22.3 Summary

In this chapter we learned how to control the constraints for one random variable depending on the value of another. We saw that we can use the value of one variable to imply the value for another, and we saw that we can use an `if` statement to turn constraints on and off.

We also got a glimpse into how the constraint solver works. We saw that it carves up the space of possible results and then randomly picks from the resulting set, rather than looking at each part of each constraint sequentially. We also learned how to use the `solve...before` construct to control the order of randomization.

In our next chapter we'll learn how to use all these techniques to create random arrays.

23 Constraining Arrays

In This Chapter:

- *Randomizing Fixed-Size Arrays*
- *Randomizing Multidimensional Arrays*
- *Randomizing Dynamic Arrays*

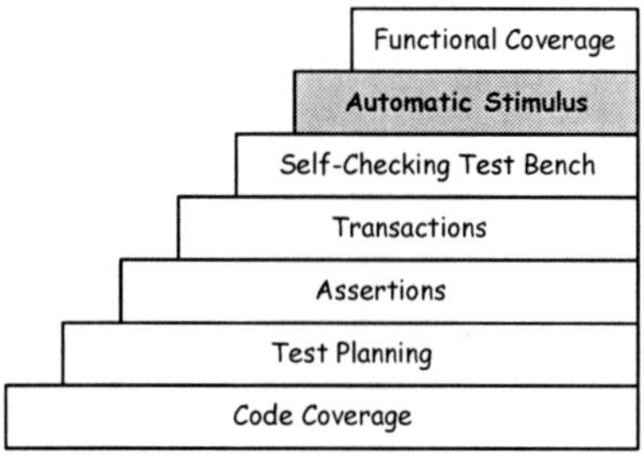

The Wishbone bus from opencores.org is an open-source interconnect bus that can be used inside of chips. The bus allows both single-word transactions and burst transactions, and the burst transactions can be any length.

If you are testing a Wishbone design, you must create a transaction that has an array to represent burst data. Because the bursts can be any length, you use dynamic arrays to represent the data.

This is why SystemVerilog allows you to randomize arrays. It gives you the ability to put arrays into transactions and randomize the elements in the array. You can also constrain each element in the array in terms of the other elements.

23.1 Randomizing Fixed-Size Arrays

Fixed-size arrays are arrays whose lengths we define at the time we declare them. SystemVerilog simply puts some memory aside when you declare the array and that is all that will ever be defined for the array.

We randomize arrays by including them in transaction objects and using the `rand` and `randc` keywords when we declare them. When we randomize an array,

SystemVerilog chooses a value for every element in the array. We identify random arrays with the `rand` keyword.

For example, let's say that we want to create a transaction that would be a frame in a color display. The display has 1024 pixels and each pixel has a color and a brightness. This object stores one frame of this display. The two arrays, respectively, store a brightness and color for each pixel (there should really be three brightnesses, one for each color, but that would be a very complicated example).

We'd define our transaction like this:

```
17  typedef enum {red, grn, blu} color_t;
18  class color_frame;
19     rand  logic [3:0] brightness [1023:0];
20     rand  color_t     color [1023:0];
21
22     task show;
23        $display ("--- frame arrays from 9 to 21---");
24        $write("index: ");
25        for (int i=9;i<=21;i++) $write("%4d",i);$display;
26        $write("logic: ");
27        for (int i=9;i<=21;i++) $write("%4d",brightness[i]);$display;
28        $write("color: ");
29        for (int i=9;i<=21;i++) $write("%4s",color[i]);$display;
30     endtask // displayarray
31  endclass
```

FIGURE 23-1. Declaring Randomized Arrays

- **Line 19**—Create an array of four-bit values that is 1 K long.
- **Line 20**—Create an array of `color_t` values that is 1 K long.
- **Line 22**—Define a method that prints out part of the arrays

Now we can randomize this transaction. Without constraints, SystemVerilog gives each entry in the arrays a random number. The problem is, that's too random. It would just produce snow. For a better test, we want to constrain the numbers that go into the array.

SystemVerilog lets us constrain the values in an array. We can constrain them in terms of constants, or of their index, or even in terms of other members of the array. SystemVerilog does this by means of the `foreach` keyword.

The `foreach` constraint looks like this:

```
foreach(array_name[loop variables]){
   constraint1;
   <constraint2>;}
```

We constrain the array by using its name in the `foreach` loop and declaring a variable to hold the index. The `foreach` loop has a block of constraints that use the index and the arrays to control how we randomize our array.

As an example, let's say that we don't want snow on our video screen. Instead, we'd like one-pixel bands of color that go from being dim to being bright. We can implement this by describing the following two rules:

1. Every pixel is the same color as the one before it and is brighter than the one before it. The exception is every tenth pixel.
2. Every tenth pixel has a brightness of zero and unconstrained color.

These rules give us bands of color ten pixels long that range from dim to bright. Each band can be red, green, or blue.

We need to constrain our `color_frame` transaction to make this happen. We'll do that with the `foreach` keyword. We'll pass the `foreach` keyword three constraints:

- The first constraint makes every tenth pixel have a brightness of zero.
- The second constraint makes every other pixel be brighter than the one before it.
- The last constraint forces the pixels to be the same color, except for pixels whose index is divisible by ten. These pixels can be any color.

Here's what that constraint looks like in a program:

```
50 module top;
51    color_frame frame;
52
53    initial begin
54       frame = new();
55       for (int i = 1; i<=3; i++) begin
56          assert(frame.randomize() with {
57           foreach (brightness[x]) {
58              if ((x%10) == 0) brightness[x] == 0;
59              if ((x%10) != 0) brightness[x] > brightness[x-1];
60              if ((x%10) != 0) color[x] == color[x-1];
61              }}
62          );
63          frame.show;
64       end
65    end
66 endmodule
```

FIGURE 23-2. Constraining Array Elements

- **Line 54**—Make a new `frame` transaction.
- **Line 56**—Randomize `frame` with the following constraints.
- **Line 57**—Use `foreach` to randomize each element in the arrays.
- **Line 58**—We'll use a conditional constraint. If the index (`x`) is divisible by 10 (`%` does modulo division), then we set the brightness to zero. Otherwise this constraint is ignored.
- **Line 59**—The second conditional constraint that applies only for indices not divisible by 10. This constraint says that the pixel in location `x` must be brighter than the one before it. It doesn't say how much brighter.
- **Line 60**—The last constraint handles `color`. If the index is not divisible by 10, then each element must be the same as the one before it. The entries in the `color` array that are not divisible by 10 are not constrained. This gives us 10 entry strings of the same color.
- **Line 63**—We created a `show()` method in the transaction to print out part of its values. We use it here.

It's important to note that these constraints do not actually put the values into the arrays. Those are being entered randomly. The constraints define a rule for each entry in the array that controls how that entry is randomized. We get different values in the arrays each time we randomize the transaction.

The code randomizes `frame` three times and prints out the results:

```
# --- frame arrays from 9 to 21---
# index:    9  10  11  12  13  14  15  16  17  18  19  20  21
# brght:   14   0   1   5   6   7   8  10  11  13  14   0   1
# color:  blu grn grn grn grn grn grn grn grn grn grn red red
# --- frame arrays from 9 to 21---
# index:    9  10  11  12  13  14  15  16  17  18  19  20  21
# brght:   15   0   2   3   4   7   9  12  13  14  15   0   1
# color:  grn red red red red red red red red red red grn grn
# --- frame arrays from 9 to 21---
# index:    9  10  11  12  13  14  15  16  17  18  19  20  21
# brght:   15   0   1   2   3   7   8   9  13  14  15   0   1
# color:  blu blu blu blu blu blu blu blu blu blu blu grn grn
```

FIGURE 23-3. A Constrained Array

These arrays have different values but follow all the constraint rules. Look at the 10th and 20th `brightness` array element and you'll see that they are always 0. The rest of the elements are all greater than the ones to their left, but the numbers increase in value differently. Finally, the `color` array is the same until the 10th or 20th element (except in the last case, where `blu` was chosen again randomly).

23.2 Randomizing Multidimensional Arrays

One-dimensional arrays are great for simulating RAM, but the real world is in three dimensions or even more. We need to be able to randomize arrays of any size.

SystemVerilog's `foreach` statement supports multiple dimensions. You add dimensions to the `foreach` by using a comma:

```
foreach(<array>[<x>,<y>,<z>,...])
```

When SystemVerilog sees a multidimensional `foreach`, it loops over all the numbers in the array range. As it loops, the left-most number changes most quickly. The variables in the `foreach` match up with the variables in the declaration.

Consequently, if we define an array to be `my_2d[1:3][1:4]` and we write

`foreach(my_2d[x,y])`

the `x` corresponds to the `[1:3]` range and the `y` corresponds to the `[1:4]` range.

Here's an example of constraining `my_2d` so that every cell is greater than the one to its left and the one above it:

```
29              assert(t.randomize with {
30                 foreach(my_2d[x,y]) {
31                    if (x>1) my_2d[x][y] > my_2d[x-1][y];
32                    if (y>1) my_2d[x][y] > my_2d[x][y-1];
33                    if (y==1 && x > 1)
34                          my_2d[x][y] > my_2d[x-1][`Y_MAX];
35                 }});
```

FIGURE 23-4. Two-Dimensional Constraint

- **Line 30**—Loops over the dimensions in `my_2d`.
- **Line 31**—Makes sure that each cell is larger than the one above it. Notice that the condition acts as a guard. It only subtracts 1 from `x` if `x` is greater than 1.
- **Line 32**—Makes sure that each cell is larger than the one to its left.
- **Lines 33–34**—If we have a cell against the left edge of the array and it is not at the top of the array, then make sure the cell is larger than the last cell of the previous row.

When we randomize this array twice, we get the following:

```
#        1  2  3  4
#       -- -- -- --
# 1:     1  3  8 13
# 2:    16 19 20 26
# 3:    27 28 29 30
#
#        1  2  3  4
#       -- -- -- --
# 1:     5  9 10 11
# 2:    15 17 20 22
# 3:    23 25 28 29
```

FIGURE 23-5. Results of Randomization with `foreach`

The two-dimensional array increases from left to right and from top to bottom. The link between rows is the constraint on lines 33 and 34. It makes sure that each row is larger than the one above it.

23.3 Randomizing Dynamic Arrays

At the beginning of this chapter we talked about the Wishbone bus and how it could have a burst transfer of any length. There is only one way to model that in SystemVerilog—with a dynamic array.

SystemVerilog can randomly create dynamic arrays of any length as long as we constrain that length. While a Wishbone bus transfer can theoretically have an infinite length, there are practical limits in any system. Therefore, we need to be able to constrain the maximum and minimum length for the array.

Dynamic arrays in SystemVerilog have a built-in function called `size()`. Normally we use this function when we want to read the size of a dynamic array, but we can also use `size()` in a constraint.

This idea of using the same word for two different operations is called *overloading*. For example, the SystemVerilog LRM overloaded the `foreach` keyword so that it can work both as a statement in procedural code and as a constraint. The same is true of the `size()` method. The LRM overloaded it so that it can either return the size of a dynamic array or constrain the array size.

When a programming language overloads a word, we use context to determine the intention of the word. This is the same as in English. For example the word "pass" is overloaded in these sentences:

> *"He made a pass at her in a bar."*
>
> *"He made a pass to her on the soccer field."*

The word "pass" has different meanings depending upon on the context. In the same way, the `foreach` and `size` keywords have different meaning depending upon on their context.

Here is an example of using `foreach` and `size` in both contexts:

```
16 class array_holder;
17    rand logic [3:0] array [];
18
19    task show;
20       $display ("--- array.size() = %0d---",array.size());
21       $write("index:\t");
22       foreach(array[i]) $write("%6d",i);$display;
23       $write("logic:\t");
24       foreach(array[i]) $write("%6d",array[i]);$display;
25    endtask // show
26
27 endclass
28
29 module top;
30   array_holder ar;
31
32   initial begin
33      ar = new();
34      for (int i = 1; i<=3; i++) begin
35         assert(ar.randomize() with {
36           array.size() > 0;
37           array.size() < 10;
38           foreach (array[x])
39              if (x>0) array[x] > array[x-1];
40           });
41         ar.show;
42      end
43   end
44 endmodule
```

FIGURE 23-6. Randomizing a Dynamic Array

- **Line 16**—Create an object called `array_holder` that—wait for it—holds an array.
- **Line 17**—We declare `array` to be a dynamic array of four-bit values.
- **Lines 19**—The `show()` method prints `array` to the screen.
- **Line 20**—The `show()` method uses the `size()` function to find the number of entries in `array`. This is the procedural context of `size()`.
- **Lines 22–24**—Use the procedural context of the `foreach` keyword to step through the array and print its indices and values.
- **Line 35**—Randomize `array` with three constraints.
- **Lines 36 & 37**—Use `size()` in the constraint context to say that the array must be at least one value long and fewer than 10 values long.
- **Lines 38 & 39**—Use the `foreach` keyword in the constraint context to loop through all the values in the array and assign a constraint to each one.
- **Line 41**—Execute the `show()` method we created on line 19.

When we run this example, we get three arrays with different lengths:

```
# --- array.size() = 9---
# index:             0    1    2    3    4    5    6    7    8
# logic:             4    7    9   10   11   12   13   14   15
# --- array.size() = 8---
# index:             0    1    2    3    4    5    6    7
# logic:             0    4    7    8    9   11   14   15
# --- array.size() = 6---
# index:             0    1    2    3    4    5
# logic:             5    8   10   12   14   15
```

FIGURE 23-7. Randomly Created Dynamic Arrays

Some readers may have noticed a problem with the constraints that made these arrays. In Chapter 22.1.2 on page 275, I said that SystemVerilog solves all the constraints simultaneously unless we use a `solve...before` construct. There are no `solve...before` lines here, so how does the `foreach` know the length of the array?

SystemVerilog addresses this issue by implicitly solving all `size()` constraints before solving any others. We didn't write the `solve...before` lines, but they were there as soon as we constrained the size.

23.4 Summary

In this chapter we learned how to constrain arrays of numbers. We learned about the `foreach` constraint and saw how to use it to constrain arrays with any number of dimensions. We also saw how powerful conditional constraints can be and how they allow us to constrain array cells in terms of other cells.

We also learned how to constrain dynamic arrays, first by constraining their size, and then by constraining their individual cells. We can now create transactions that represent burst operations.

We've now completed our discussion of how to create constraints, so we can move on to discussing where to put them. Until now we've attached constraints to a call to `randomize()` by using the `with` keyword. This usage is easy to understand and provides all the constraint information at a logical point in the source code. But it's not the only, or necessarily the best, place to put a constraint.

It is possible to include constraints inside class definitions. This allows us to capture innate constraints so that we don't need to repeat them whenever we randomize.

In our next chapter, we'll learn how to put constraints into an class definition, and in the last chapter on automatic stimulus, we'll create the test bench that writes its own tests.

24 Using Constraints in Objects

IN THIS CHAPTER:

- *Adding Constraints to Objects*
- *Extending Objects*
- *Controlling Constraints in Objects*

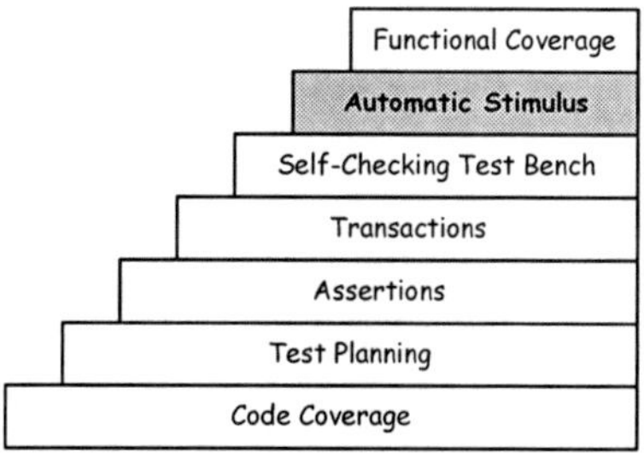

Our goal, as always, is to create a test bench that creates its own stimulus by using constrained random numbers. We've seen how to use the `with` statement to control calls to the `randomize()` method. This is a powerful way of adding constraints to a design, but there is another way. Rather than tying the constraint to the `randomize()` call, you can make it part of an object.

The `with` statement is nice because it clearly connects the constraints associated with a randomization to the randomization call. You can see all the randomizing information in one place. There is, unfortunately, a downside to this approach.

There are many times when a constraint is actually part of the transaction rather than a testing tool. For example, let's say we have a transaction that always reads a memory on a word boundary, and let's say this transaction has an address in it. If that address is going to access memory on a word boundary, then the least significant bit must always be zero. This is a constraint on randomization.

Now, we could implement that constraint using the `with` statement whenever we make a new transaction. This is error prone. Some future designer may forget to add the constraint and spend a lot of time generating bad data. It would be better simply to associate the "least significant bit is zero" constraint with the transaction.

SystemVerilog allows you to do this. It allows you to add a constraint block to an object definition and, thus, make the constraint an innate part of the object. Once that is done, all randomization on that object automatically implements the innate constraint.

In this chapter, we add constraints to an object. Then, we look at how to define classes in terms of other classes in order to make it easier to add constraints. Finally, we'll put it together and show how to use both techniques to create a family of transactions.

24.1 Adding Constraints to Objects

We declare a constraint block to be part of the object in the same way that we declare data members. The constraint blocks sit outside all the methods. It has the following syntax:

```
constraint <constraint name> {
    <constraint1>;
    <constraint2>;
}
```

Notice the odd syntax quirk. There is no semicolon after a constraint block, even though there are semicolons after all the constraints.

Here is an example of a constraint in an object:

```
class big_number;
   rand logic [8:0] num;

   constraint make_it_big {
      num > 128;
   }
endclass
```

FIGURE 24-1. Constraining an Object

This is a great example of an object that benefits from having the constraint defined in it. If we didn't have this constraint in the object, then the programmer who randomized the object would always have to remember to add the "`num>128;`" constraint to the `with` statement. Now, anyone who randomizes a variable of type `big_number` automatically gets a number greater than 128.

These are the basics of declaring a constraint in an object. This approach allows us to encapsulate the randomization behavior of an object inside the object. The approach gets more powerful when we extend objects.

24.2 Creating Families of Classes

In Chapter 11, we introduced the concept of SystemVerilog classes and we used classes to define objects. We said that classes are really just fancy records that can hold both data and procedures or functions. We saw how to define a class and then how to use the `new()` method to create new objects based on that class.

Now we'll add another aspect to the concept of classes: *extending* a class. When we extend a class, we create a new class that is based on an existing class. Extending a class allows us to create special cases of our transactions without having to copy all the code again and again. It also allows us to make changes in one place in our code and have them propagate throughout the class family.

We extend a class by defining a new class and using the `extends` keyword to designate the parent class, as in Figure 24-3 on page 294.

Here is a concrete example of extending a class. Let's say we have a transaction called `memory_read`. This transaction contains the data from reading a 16-bit memory. Let's also say we have a transaction called a cache_read. This transaction is just like the `memory_read`, but it has a bit called `cache_hit` that tells us whether we had a cache hit on the read.

A `cache_read` is obviously just a `memory_read` with a bit added to it. We we talk about objects with a relationship like this, we say that the `cache_read` class *extends* the `memory_read` class.

An extension like this can be conveyed graphically by tools like the Mentor Graphics application called Certe:

FIGURE 24-2. Cache Read Extends Memory Read in Certe

The above shows that a `cache_read` is a kind of `memory_read`. The `cache_read` transaction extends the `memory_read` transaction.

SystemVerilog includes the keyword `extends` to allow you to define this kind of relationship. Here's how we define `memory_read` and `cache_read` and use `cache_read` in a module:

```
1  class memory_read;
2    logic [15:0] read_data;
3  endclass
4
5  class cache_read extends memory_read;
6    bit cache_hit;
7  endclass
8
9  module top;
10   cache_read c;
11
12   initial begin
13     c = new();
14     c.read_data = 'hF5;
15     c.cache_hit = 1'b1;
16   end
17
18 endmodule
```

FIGURE 24-3. Extending a Class

- **Lines 1–3**—Define a class called `memory_read` that holds a 16-bit number called `read_data`.
- **Lines 5–7**—Define a class called `cache_read` that extends `memory_read`. This means that `cache_read` inherits `read_data`. We add a bit called `cache_hit` to `cache_read`.
- **Line 10**—Declare a `cache_read` handle.
- **Line 13**—Create a new `cache_read`.

- **Line 14**—Put data into `read_data` (inherited from `memory_read`).
- **Line 15**—Put data into `cache_hit`.

The concept of extending a class is very powerful, and there are hundreds of books written on the topic of object-oriented programming. That said, we now have enough information about extending objects to add constraints to our transactions and, in the next chapter, to create a test bench that writes its own tests.

24.3 Extending Objects with Constraints

In Section 24.1 we saw how to add constraints to an object. In Section 24.2 we saw how to create families of classes. Now, we put the two concepts together and extend objects by adding constraints to them.

There are two reasons why we would choose to add constraints to an object rather than use the `with` statement:

1. The constraint may be inherent to the definition of the object. We'll see this case in this section.
2. It may simply be easier to write the test bench if we take this approach. We'll see that case in the next chapter.

This example shows how we can take a simple—unconstrained—class and make it more powerful by extending it and adding constraints. Our goal is to create the following family of classes:

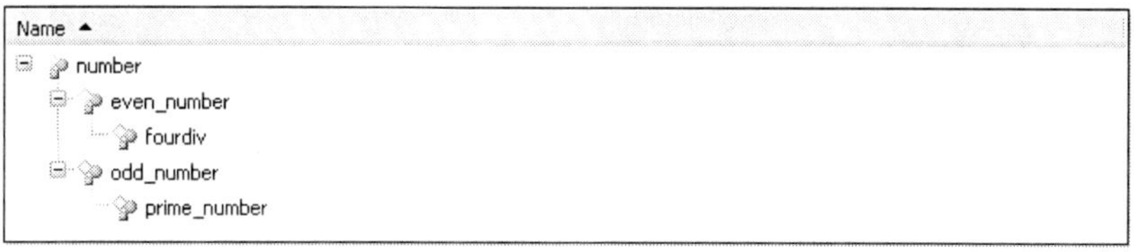

FIGURE 24-4. Number Inheritance

We can use this family of classes to create five different kinds of eight-bit numbers. The base class, `number`, is a simple eight-bit number with no constraints.

When we call `randomize()` on `number`, we can get any number between 0 and 255. The other classes have constraints that force `randomize()` to create certain kinds of numbers:

- `number`—Just an eight-bit number.
- `even_number`—An eight-bit number ending in zero.
- `odd_number`—An eight-bit number ending in one.
- `fourdiv`—An eight-bit number whose bottom two bits are zero.
- `prime_number`—An odd number that is also prime.

Figure 24-4 on page 295 shows that these classes are all related to each other and extend from the basic number class. They all have an eight-bit data member and they have no methods.

The only difference between these numbers is the way they act when we randomize them. We control this randomization with constraints.

Here's how we make this inheritance diagram a reality:

```
1  class number;
2    rand logic [7:0] num;
3  endclass
4
5  class even_number extends number;
6    constraint zerobit {num[0] == 0;}
7  endclass
8
9  class odd_number extends number;
10   constraint oddbit {num[0] == 1;}
11 endclass
12
13 class fourdiv extends even_number;
14   constraint zerobit1 {num[1] == 0;}
15 endclass
16
17 class prime_number extends odd_number;
18   constraint prime_number {
19       num inside {2,3,5,7,11,13,17,19,23,29,31,37,41,43,47,53,
20       59,61,67,71,73,79,83,89,97,101,103,107,109,113,127,131,137,
21       139,149,151,157,163,167,173,179,181,191,193,197,199,211,223,
22       227,229,233,239,241,251};
23              }
24 endclass
```

FIGURE 24-5. Implementing an Inheritance Hierarchy

- **Lines 1–3**—The class `number` is the base class that defines our eight-bit number called `num`.
- **Lines 5–7**—An `even_number` is a `number` whose least-significant bit is 0. We extend `number` and add the `zerobit` constraint to make this happen.

- **Lines 9–11**—Conversely, an `odd_number` is a `number` whose least-significant bit is a 1. We extend `number` and add the `oddbit` constraint.
- **Lines 13–15**—The `fourdiv` is an `even_number` whose bottom two bits are 0. We've extended `even_number` and added a constraint called `zerobit1`. Because we've extended `even_number,` the `zerobit` constraint from that class still applies. We now have two constraints on `fourdiv` objects.
- **Lines 17–24**—A `prime_number` is an `odd_number` that is one of a set of values. We went to Google and found a list of prime numbers; then we cut and pasted the list into an `inside` constraint. It might have been possible to create a complex set of constraints to make the simulator find the prime numbers, but this approach runs faster. *Always give the constraint solver as much help as possible.*

We'll declare all these objects, instantiate them, and then randomize them:

```
module top;
   number        n = new;
   even_number   e = new;
   odd_number    o = new;
   fourdiv       f = new;
   prime_number  p = new;

   initial begin
      $display("Three sets of Numbers...");
      repeat (3) begin
        assert(n.randomize());
        assert(e.randomize());
        assert(o.randomize());
        assert(f.randomize());
        assert(p.randomize());

        $display("n: %3d,  e: %3d, o: %3d, f: %3d, p: %3d",
                  n.num, e.num, o.num, f.num, p.num);
      end
   end
endmodule
```

FIGURE 24-6. Using Our `number` Objects

Our `randomize()` calls have gotten simple again. Because the constraints are stored in the objects, we don't have to worry about constraining the randomization. The class definitions took care of that for us.

This code also has the advantage of making logical sense. When we declare `p` to be a `prime_number`, we expect that it will give us prime numbers when we randomize it.

Here are our randomized numbers:

```
# Three sets of Numbers...
# n:   0,  e: 110, o: 249, f:  44, p:  47
# n: 124,  e:  40, o: 191, f:  16, p: 131
# n: 134,  e:  46, o: 143, f: 212, p:  59
```

FIGURE 24-7. Randomizing our Hierarchy

All the numbers follow the expected rules.

24.4 Summary

In this chapter, we learned how to add constraints to an object to make them part of a transaction. This is especially useful when the transaction has an inherent property that must always be respected when we randomize it. For example, an odd number must always end in 1.

Then we learned how to create families of classes with the `extend` statement. This allows us to create new objects by leveraging existing objects. Extending objects also makes it easier to change our objects, because changes we make to the base object ripple through the family.

Finally, we put the two concepts together and created a family of classes by adding constraints to objects. We saw that the constraints are cumulative. The extended objects use the constraints of the base object.

Now we are ready for our final chapter on automatic stimulus. We create a test bench that generates its own stimulus based on constraints that we provide. We see that when we stop writing tests by hand and start defining them with constraints, we can create powerful test benches quickly.

25 The Constrained Random Test Bench

IN THIS CHAPTER:

- *Using the Test Assembly Line*
- *The* `tester` *Module*
- *The* `run.do` *Script*

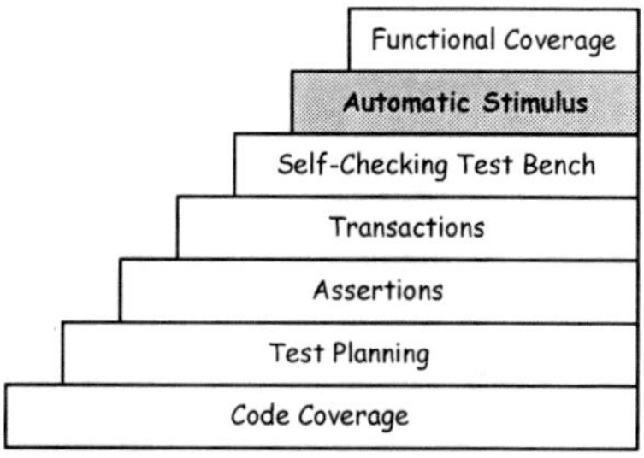

Most engineers create simulation tests the way a craftsman creates a chair: slowly and with great care. This is not what we need to simulate our designs quickly and completely. We need our tests to be cranked off an assembly line staffed by robots.

The robots are controlled by programs. In a perfect world, we just give the test bench a set of robot programs and let it create the tests. We create just such a robot-driven test bench in this chapter. The test bench is easy to write and easy to use. It cranks out tests and drives a simulation towards 100% code coverage.

We start by understanding how this robot-driven assembly line works. Then we write the assembly line in SystemVerilog. Our example, as always, is our friend the TinyALU.[1]

25.1 Using the Test Bench

First, a bit of terminology. The industry calls this kind of test bench a *constrained random* test bench, and that's what we call it in the rest of the book. Constrained

1. There are other examples on www.fpgasimulation.com.

random test benches create randomized transactions and drive them through a design. The engineer who uses these test benches controls them by designing the constraints and letting the constraints do the work.

Our constrained random test bench contains a folder called `tests`. We control the test bench by writing tests and putting them into this folder. The test bench reads each test and uses it to create 1000 transactions that fulfill the test—much like assembly line robots reading their program.

So what are these tests? Constraints! We've been learning how to create constraints, now we are going to use them to enhance our test bench. Here is a constraint that tests the TinyALU's ability to add:

```
constraint test {
  op == add_op;}

```

FIGURE 25-1. The Test File `add.svh`

This file is a test in our test bench. In this case it is an add test, because it limits the `alu_operation` transaction to `add_op`. We will get 1000 additions with randomized numbers.

This is just one of many files in the `tests` folder. Each time we want to write a test, we simply create a new file containing a constraint and put it into the folder. We call these files *tests*:

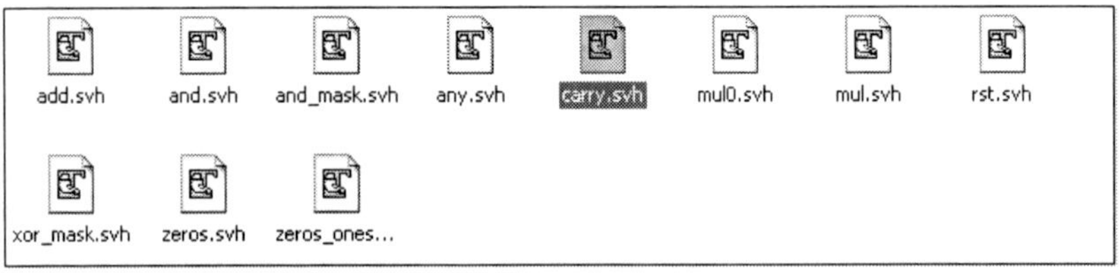

FIGURE 25-2. The `tests` Folder

The test bench goes through this folder and reads each test. It then uses the test to create 1000 transactions. As another example, let's look at the `carry.svh` file (selected in the picture above). This test makes sure that the TinyALU is handling carries properly. Here is the constraint:

```
constraint test {
  op == add_op;
  A+B >= 16'h100;}
```

FIGURE 25-3. The Carry Test: `carry.svh`

The test bench uses this constraint to create 1000 transactions that generate a carry. We define all our tests by creating constraint files and adding them to the `tests` folder.

In this chapter, we implement a test bench that reads these constraint files and runs them through our DUT.

This test bench needs to be a self-checking test bench, because it runs thousands of automatically generated transactions through the DUT. We created a self-checking test bench in Chapter 17. Here is that test bench:

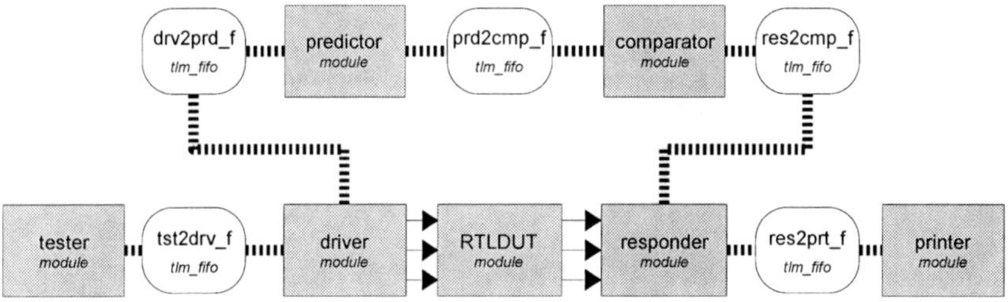

FIGURE 25-4. Our Transaction-Level RTL Test Bench

Our new test bench looks just like the self-checking test bench, except that we are going to rewrite the `tester` module. The old `tester` module (Figure 15-2 on page 183) could create specific transactions, or randomize transactions without using constraints.

The new `tester` module uses the constraint files from the `tests` folder.

25.2 The `tester` Module

The Holy Grail of test writing is to write one file and have that file describe thousands of new transactions. We don't want that one file to be a test-bench source file, because constantly editing our source files would invite errors. Instead, we want a file that sits outside the test bench, but that controls the test bench's behavior.

We implement this Holy Grail in the `tester` module. The new `tester` module uses the constraint files in the `tests` folder (Figure 25-2 on page 300) to create thousands of transactions.

The `tester` uses two features of SystemVerilog:

- **Include files**[2]—SystemVerilog has the ability to read a file from the disk and include it in the source code, so that it gets compiled as if it had been written into the code.
- **Class extension**—In Chapter 24 we learned how to extend a class to create a family of classes. We saw that we could add constraints when we extended a class and that the new class would have the properties of the original class plus the constraints.

The new `tester` module source file contains a class definition that extends the basic transaction class (in this case, `alu_operation`), and adds a constraint to it. The constraint comes from one of the constraint files in `tests` folder. Here is the source code that does this in our `tester` module:

```
import ovm_pkg::*;
import tinyalu_pkg::*;

class test_request extends alu_operation;

  `include "test_constraint.svh"

endclass

module alu_tester(ref tlm_fifo #(alu_operation) op_f);

test_request op;
```

FIGURE 25-5. Class Extension in the `tester` Module

2. Discussed in Section 6.5 on page 64.

- **Lines 16–17**—As always, we import `ovm_pkg` and `tinyalu_pkg`.
- **Line 19**—We create a new class, called `test_request`, which extends the `alu_operation` class. It is an `alu_operation` with an additional constraint.
- **Line 21**—We include the file `test_constraint.svh`, which contains the additional constraints for this test. As we'll see below, this file was copied from the `tests` folder and contains a constraint.
- **Line 27**—We declare the variable `op` to be of type `test_request` (not of type `alu_operation`). Now when we randomize `op`, we'll get the constraint we want.

At this point we have a new class, called `test_request`, that extends the `alu_operation` class. This `test_request` class has the randomization constraints that we want to use in this test, because it read them from `test_constraint.svh`, the origins of which we'll cover in a few pages.

Here is the rest of the `tester` module:

```
28
29  string s, m;
30    task run;
31       op = new();
32       repeat (1000) begin
33          assert(op.randomize());
34          $sformat(s,"Request: %s",op.convert2string);
35          $sformat(m,"%m");
36          ovm_top.ovm_report_info(m,s,300);
37          op_f.put(op.clone());
38       end
39       #500;
40       ovm_top.die();
41    endtask // run
42  endmodule
```

FIGURE 25-6. Generating Transactions

- **Line 30**—The `tester` is a transaction-level module, so it has a `run()` task.
- **Line 31**—Create a new transaction called `op` of type `test_request`. This transaction contains the constraints we want.
- **Line 32**—Create 1000 transactions.
- **Line 33**—Randomize `op`. We don't need a `with` statement, because the constraints are inside the transaction.
- **Line 36**—Print the transaction to the screen with a verbosity of 300.
- **Line 37**—Clone the transaction and put it into `op_f`.

Now we have a module that creates transactions that match the constraints stored in the `test_constraint.svh` file.

The new `tester` module is the only difference between the constrained random test bench we describe here, the self-checking test bench we described in Chapter 17 and the transaction-level test bench we described in Chapter 15. The rest of the SystemVerilog source files are exactly the same.

25.3 The `run.do` Script

We've seen that we have a folder call `tests` that is filled with test files (see Figure 25-2 on page 300). We've also seen that the `tester` module expects to include a file called `test_constraints.svh`. Now we'll see how we copy each test file from the `tests` folder and put it into the top-level folder as the `test_constraints.svh` file.

Once you start creating advanced test benches, the script you use to run the tests becomes an integral part of the test bench. In our case, the `run.do` script handles the job of simulating the DUT with every set of constraints in the `tests` folder.

This step in the process requires scripting in the Tcl language. Tcl stands for Tool Command Language and was created by John Ousterhout. Tcl has become a common command language across most design tools, and it is also common to the major simulators.[3]

Tcl scripts connect the tools to the file system and pull everything together. In our case, our `run.do` script does the following:

1. Compiles the test bench and DUT, except for the `alu_tester.sv` file.
2. Loops over all the files in the `tests` folder.
3. Copies each file to `test_constraints.svh`.
4. Recompiles `alu_tester.sv` with the new `test_constraints.svh` file.
5. Runs the simulation with the new `tester` module and stores the results in the `debug` folder.

3. There are many books on Tcl, but I still like the original: Tcl and the Tk Toolkit by John K. Ousterhout.

This system gives us a separate simulation run for every constraint file in the tests folder. Because there are 11 constraint files and 1,000 transactions per file, we'll see 11,000 transactions go through the TinyALU.

Because all simulators support Tcl, they all handle file names, loops, and lists in the same way. The only difference between the simulators is that they have different command names for compilation and simulation.

Here is the `run.do` script. It was written to work in Questa from Mentor Graphics, but it also works with the other simulators as long as you change the names of the simulator commands.

We don't want to compile the entire test bench each time we run a test. Instead, we recompile only one file for each test: `alu_tester.sv`. Therefore, we compile all the other files in the test bench first. Then we compile `alu_tester.sv` and run the simulator once for every file in the `tests` folder.[4]

Here is the `run.do` script that runs our test bench. We run thousands of transactions through the test bench by running this script once:

```
if [file exists "work"] {vdel -all}
vlib work
vlog -f compile_sv.f
vcom -f compile_vhdl.f
set contents [glob -directory tests *]
foreach item $contents {
        file copy -force $item ./test_constraint.svh
        vlog alu_tester.sv
        puts "********************************************"
        puts "********************************************"
        puts "                    $item"
        puts "********************************************"
        puts "********************************************"
        set testpath [split $item "/"]
        set testfile [split [lindex $testpath 1] "."]
        set testname [lindex $testfile 0]
        vsim top -novopt -coverage -debugDB=debug/$testname.dbg -wlf deb
        set NoQuitOnFinish 1
        onbreak {resume}
        log -r /*
        run -all
}
```

FIGURE 25-7. Tcl Test Bench Control Script

4. You can ask questions about Tcl on the forums at www.fpgasimulation.com.

- **Lines 1–4**—These Questa commands compile the device under test and the test bench, except for the `alu_tester.sv` file.
- **Line 5**– This Tcl line reads the file names out of the `tests` folder and stores them in a list called `contents`.
- **Line 6**—This loop runs once for each item in the `contents` list. Each item contains the path to a constraint file, such as `tests/add.svh`
- **Line 7**—Copy a test file to `test_constraint.svh`.
- **Line 8**—Recompile `alu_tester.sv` with the new include file.
- **Line 17**—Launch the simulation.
- **Line 21**—Run the simulation to the end. Then go back to the top of the loop (on line 6) and get another test.

This script's structure works for any design. It allows you to create new tests by writing another constraint file and dropping it into the `tests` folder. You can easily create thousands, or millions, of transactions this way.

25.4 Examining the Output

Creating thousands or millions of transactions is a great way to achieve 100% code coverage, but it also creates a tremendous amount of output. We can cope with that output in two ways:

1. Use the `ovm_top` object to limit the number of messages we see.
2. Store our output in files so we can analyze it later.

By limiting the output, and storing the output we have in a set of files, we can manage all this data.

25.4.1 Limiting Simulation Output

The Open Verilog Methodology (OVM) allows us to limit output in several ways:

- Limit the number of messages with the verbosity level.
- Disable messages from uninteresting message IDs.
- Stop the simulation after a small number of errors.

Here are the OVM calls that limit the output from our constrained random test bench. They demonstrate all three approaches:

```
53   initial begin
54         ovm_top.set_report_verbosity_level(200);
55         ovm_top.set_report_max_quit_count(2);
56         ovm_top.set_report_id_action(`OUTPUTID, OVM_NO_ACTION);
```

FIGURE 25-8. Limiting Output with the OVM

- **Line 54**—Allow only error messages to be printed. All of our informational messages have a verbosity of 300 (see Figure 25-6 on page 303), so they don't show up.
- **Line 55**—Stop the simulation if there are two errors. There's no need to keep going.
- **Line 56**—Don't print the output from the printer module.

Now we've limited the amount of data that we'll produce—but we still need to store that data in a central location so we can analyze it. We use the `debug` folder for that.

25.4.2 The `debug` Folder

When we ran the simulations, we told the simulator to store the transcript files in a folder called `debug`. The kinds of files we store vary according to the simulator used.

Running this design on my machine gave me the following debug folder:

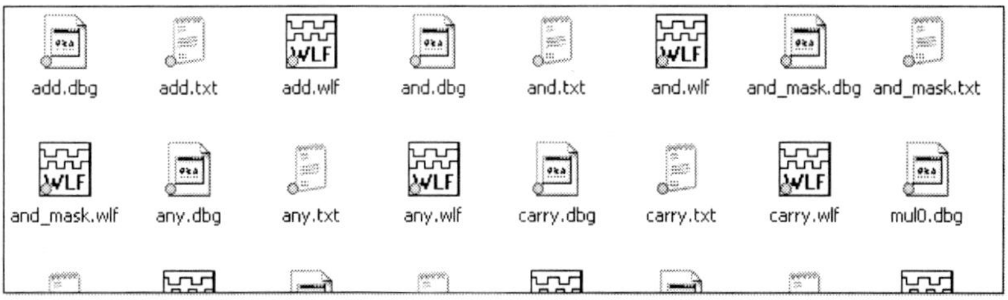

FIGURE 25-9. The `debug` Folder with Questa Debug Files

Every file in the `debug` folder corresponds to a file in the `tests` folder. Now we use a searching tool such as `grep`[5] to find files that contain the word "`OVM_ERROR`":

```
1  add.txt:# OVM_ERROR :     0
2  and.txt:# OVM_ERROR :     0
3  and_mask.txt:# OVM_ERROR :     0
4  any.txt:# OVM_ERROR @ 370 [top.comparator.run]
5  any.txt:# OVM_ERROR @ 810 [top.comparator.run]
6  any.txt:# OVM_ERROR :     2
7  carry.txt:# OVM_ERROR :     0
8  mul.txt:# OVM_ERROR :     0
9  mul0.txt:# OVM_ERROR :     0
10 rst.txt:# OVM_ERROR @ 2490 [top.comparator.run]
11 rst.txt:# OVM_ERROR @ 3190 [top.comparator.run]
12 rst.txt:# OVM_ERROR :     2
13 xor_mask.txt:# OVM_ERROR @ 50 [top.comparator.run]
14 xor_mask.txt:# OVM_ERROR @ 90 [top.comparator.run]
15 xor_mask.txt:# OVM_ERROR :     2
```

FIGURE 25-10. Results of Searching for `OVM_ERROR`

- **Lines 4–5**—Errors in the `any.svh` test.
- **Lines 8–9**—Errors in the `rst.svh` test.
- **Lines 13–14**—Errors in `xor_mast.svh` test.

These errors are caused by the same bug that was in Chapter 17.[6]

25.5 Summary

In this chapter, we learned how to create a robot-driven assembly line that cranks out tests based on our instructions. We saw that these instructions are randomization constraints, and that we could create a script and a test bench that would create and test transactions that were created with those constraints.

We learned how to extend a basic transaction class to include the constraints, and how to use SystemVerilog's ability to include a file to create a test bench that needs to compile only one file per test.

5. grep is a utility in the Unix/Linux world. A Windows version is available at www.wingrep.com.
6. The fix is left as an exercise for the reader—or one could ask on www.fpgasimulation.com.

We also saw how to create a folder of OVM output, and how to search that folder to find errors.

Using a test bench such as this, we can quickly simulate most of the corner cases in our design and achieve nearly 100% code coverage. Then we can write directed tests that get us to complete 100% code coverage.

Unfortunately, it turns out that there is a problem. We have been using 100% code coverage as our Holy Grail. Our assumption has been that if we simulate every line in a design, then we'll have tested all the functionality in the design. As we'll see in our next chapter, this is not necessarily the case.

Code coverage assumes there is a link between the code in our design and the actual functionality we are testing. This link is not as solid as one might think. Our final step is Step Seven: Functional Coverage.

26 Introduction to Functional Coverage

IN THIS CHAPTER:

- *What is Functional Coverage?*
- *Coverage-Driven Testing*
- *Understanding Coverpoints*
- *Implementing Coverpoints*

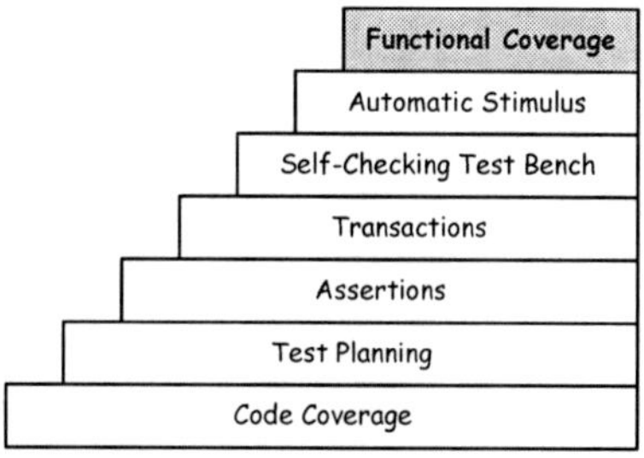

We've traveled a long way since our very first step up the FPGA verification staircase. In that first step, we added code coverage to our repertoire and created a definition of complete verification:

Verification is complete when we have 100% code coverage and a reviewed list of exceptions.

The theory here is that if we didn't simulate a line of code, then we didn't test it. Is the opposite true? If we simulated a line of code, does that mean we have tested it? Consider the following example.

Let's run a test on our TinyALU example and check the coverage. Here is the statement coverage and branch coverage after we've run our test:

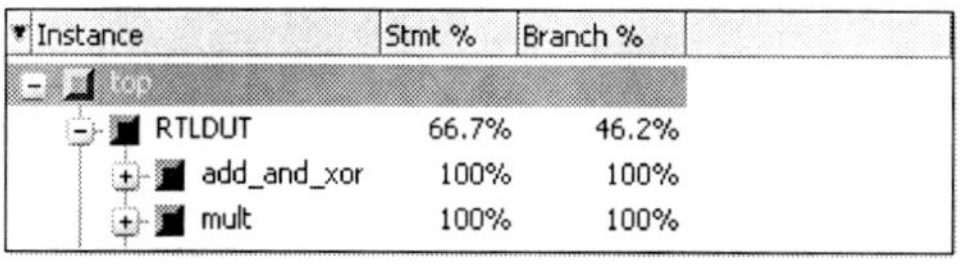
Instance
Stmt %
Branch %
top
RTLDUT
66.7%
46.2%
add_and_xor
100%
100%
mult
100%
100%

FIGURE 26-1. Statement Coverage and Branch Coverage on TinyALU[1]

This test looks pretty good! The TinyALU consists of two blocks: one handles single-cycle operations, and the other handles multiplies, which are multicycle operations. It looks like we've covered both completely. Have we?

The the top level, `RTLDUT` gives us some clue that there is a problem. While both blocks have 100% coverage, the top level does not. The problem becomes clear when we look at our test constraint:

```
constraint test { op != mul_op;  }
```

FIGURE 26-2. Single-Cycle Test Constraint

This constraint says that we never tested the `mult` block in the TinyALU. So how did the `mult` block get 100% code coverage?

This example demonstrates the problem with using code coverage as the sole tool for determining whether we have tested a design. Just because we have simulated code, it doesn't mean that we have tested the output.

The TinyALU runs all operations through both the single-cycle and multicycle blocks. It uses a MUX at the top level to decide which result reaches the output pins.

While we simulated the multiply block several times, we never tested its output. Code coverage can tell us only what we missed—not what we tested. We need functional coverage for the latter, and functional coverage is the last of the seven steps to complete FPGA simulation.

26.1 What is Functional Coverage?

I once visited a warehouse that shipped cosmetics. The warehouse received orders for cosmetics, boxed them up, and mailed them. They had a two-step process for ensuring that they shipped the right products in the right box.

1. All coverage screenshots in this and the following chapters are from Questa by Mentor Graphics

The first step happened when the box was packed: the warehouse worker compared the items in the box to the list of things on the order. The worker then made a check mark on the list for each item, to ensure that everything on the list was in the box.

The second step happened just before the box was shipped: a different worker weighed the box and compared its predicted weight (based on the items in the order) to the actual weight. If they were the same, then the box was shipped.

Functional coverage and code coverage have a similar relationship to the system in the warehouse. When we do functional coverage, we check our results against a list of behaviors we are supposed to test. We "make a check mark" next to each behavior on the list.

Code coverage is like weighing the box. If we have less than 100% code coverage, we know that we missed something. Together the two forms of coverage give us a matrix that helps us debug our tests. By seeing what kind of coverage we are missing, we know what sort of work is left:

	Missing Functional Coverage	100% Functional Coverage
Missing Code Coverage	Nothing good here. You haven't run enough tests and your test list may be incomplete.	You've tested all the functionality, but you don't have full code coverage. Either your list of functionality is incomplete, or you have dead code.
100% Code Coverage	Either your coverage list describes functionality that is not supposed to be in the design, or your design is incomplete.	You are done!

FIGURE 26-3. Implications of Missing Coverage

Combining functional coverage with code coverage gives us a complete view of our status. It's especially useful for catching dead code—code that we inherited from a previous designer but don't use. For example, we may have a piece of IP that does burst transfers, but we never use burst transfers.

26.2 Coverage-Driven Testing

Before we get into the details of implementing functional coverage, let's take a step back to see how the capabilities of functional coverage can make our verification more effective and allow us to describe our testing requirements more effectively.

Traditionally, we imagine test plans being written from a specification as a two-step process:

1. Make a list of all the pieces of functionality in the design.
2. Write the name of a test that tests each piece of functionality.

At first glance this seems like a perfectly reasonable approach. The problem is that this approach makes an assumption that can miss functionality, and it has a flaw that causes us to take much longer to verify designs than necessary.

When we adopt this "one test per function" approach to test planning, we are assuming that a test called, for example, *read_after_write* really tests the read-after-write functionality. In fact, the test may have a bug in it that causes it to miss crucial parts of the read-after-write function.

For example, in 1994 Intel lost over $400 million dollars when it allowed a bug in floating-point operations to ship in a Pentium processor. Intel had a test for the division operator (FDIV), but the test did not fully verify a lookup table that was part of the algorithm. This was a case where a test didn't really test its intended functionality. The problem would have been caught by functional coverage.[2]

The "write a test per function" approach has a second problem: it is inefficient. Remember the Wishbone bus? It has reads, writes, burst reads, and burst writes, all with different lengths and widths. It would take forever to write a test for each possible combination.

Fortunately, functional coverage gives us an alternative called a *coverage-driven test plan.* When we write a coverage-driven test plan, we don't worry about writing a test for each function. Instead, we make sure that we have a functional coverage point for each function.

2. Because the bug was in the data of the lookup table, Intel would have achieved 100% code coverage and still not seen the bug.

This approach is already being used successfully by engineers who create the custom ICs in our cell phones and computers. It has two advantages:

1. When we create coverage points for each piece of functionality, we directly measure what the tests have accomplished.
2. We don't need to write individual tests for each piece of functionality. We can use automatically generated stimulus to see what coverage we've achieved. Then we can write directed tests to fill in the blanks. This approach allows our test bench to do most of the work.

Whether we are writing a traditional test plan (list of tests) or a coverage-driven test plan (list of things to cover), we need to list each piece of functionality that must be covered. The "functionality that must be covered" is described by a new concept: coverpoints.

26.3 Understanding Coverpoints

When we use functional coverage, we capture all our functionality in a list and run our simulation. The simulator then tells us whether we've tested everything we need to test. The question remains, "A list of what?" The answer is, "A list of coverpoints."

Coverpoints are descriptions of each thing that we need to make sure we test in our design. For example, if we are to test the TinyALU completely, we need to test each operation. The fact that we missed the multiply operation in our opening example showed us that we missed a coverpoint.

However, coverpoints can be more complex than simply a list of operations. There are two types of coverpoints:

- **Signal coverpoints**—Signal coverpoints check to make sure that we've seen essential signal transitions in our design. For example, if we are testing a cache, we want to make sure we see the `cache_miss` signal go high at least once. This behavior would be captured as a signal coverpoint.
- **Data coverpoints**—Data coverpoints look at the values stored in the variables of a design and count the number of times they see specific values or combinations of values. For example, we want to make sure we test all the operations exercised by the TinyALU. Each operation is captured as a data coverpoint.

The coverage-based test plan is nothing but a long (very long in some cases) list of signal and data coverpoints. For example, here is a coverage model for the TinyALU. We want to make sure that we see the following:

- The `start` signal has been raised at least once (signal)—we watch the start signal and count the number of times it's gone high.
- The `done` signal has been raised at least once (signal)—this shows that the TinyALU has completed at least one operation.
- The `start` signal stays high for four clock cycles (signal)—this signal coverpoint ensures that we've tested a multiple-cycle operation.
- We have tested all operations (data)—this coverpoint makes sure that we've had at least one operation of each type.
- We have done all operations with all zeros and all ones (data)—this coverpoint captures a common test mode.
- And so on and so on as needed.

This list is incomplete, but it captures the idea behind a coverage-based test plan. Our goal in functional coverage is to create a complete coverage model that can be fulfilled only if we've tested the design fully.

26.4 Implementing Coverpoints

Coverpoints are nothing more than counters that say "This happened" or "I saw this data go by." In that sense, they could be implemented with any code that notices behavior and writes data to a file.

However, writing functional coverage points by using procedural SystemVerilog or VHDL would be very difficult. Because there are many details associated with capturing both signal and data coverpoints, it would be inefficient to try to capture the behavior with plain Verilog code.

SystemVerilog provides two language tools that help us write functional coverage points:

- **SystemVerilog assertions (SVA)**—SystemVerilog assertions allow us to define the behavior of signals and the relationships between them. While this ability can allow us to catch bugs, it can also be used to describe signal coverpoints.[3]

3. We are not going to talk about writing SystemVerilog assertions. They are complicated enough that they could take up an entire book. We use them by instantiating OVL checkers.

- **SystemVerilog covergroups**—Covergroups were designed specifically for functional coverage. They can capture data that has been run through the test and they can also capture signal behaviors.

Asking when you should use each of these technologies is a little like asking when you should use a tack hammer instead of a regular claw hammer. They both do roughly the same thing, but they do it differently enough that it's OK to have two different tools in the tool box.

Generally speaking, SystemVerilog assertions are better for signal coverpoints, and SystemVerilog covergroups are better for data coverpoints. In our case, we are going to simplify the process by focusing on two implementation strategies: using the OVL and using SystemVerilog covergroups.

26.5 Summary

In this chapter we learned about a new form of coverage called *functional coverage*. Functional coverage moves beyond code coverage to ensure that we've tested all the functionality in a design specification completely.

We learned that we can use functional coverage, in combination with code coverage, to make sure that we've tested a design fully. We saw that we can have 100% code coverage on a block and still not have tested its functionality completely.

Coverpoints are behaviors that we have decided we need to see in order to test a design completely. There are two kinds of coverpoints:

- Signal coverpoints describe signal behaviors that we need to see. These can be events such as seeing the `full` signal go high in a FIFO.
- Data coverpoints describe data that we need to see go through our design. For example, we may want to make sure that we touch all the addresses in a lookup table.

Coverpoints can be implemented with SystemVerilog assertions or SystemVerilog cover groups. In the next chapter, we are going to use the OVL to access SystemVerilog assertions, and we are going to learn how to write our own SystemVerilog covergroups.

27 Signal Coverpoints with the OVL

IN THIS CHAPTER

- *Using Coverpoints in the OVL*
- *Instantiating OVL Checkers for Coverage*
- *Simulating with OVL Functional Coverage*

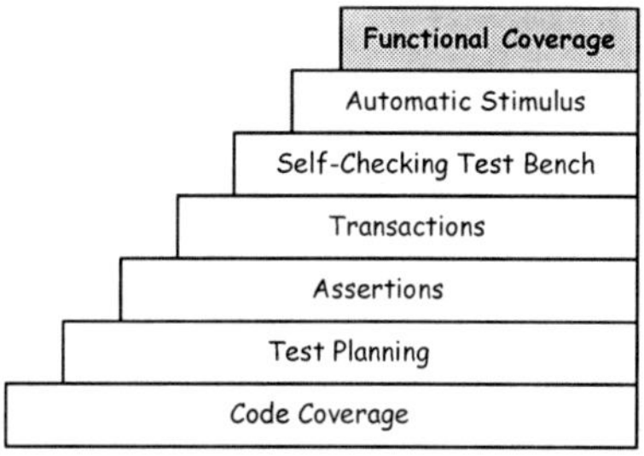

In the previous chapter, we learned about functional coverage. We saw that functional coverage allows us to describe the kinds of behaviors we want to test, and to check that they have been tested. We capture these desired behaviors with two kinds of coverpoints:

- **Signal coverpoints**—These capture behaviors associated with a signal or set of signals. For example, we may want to ensure that we tested a burst transaction by seeing that the burst signal went high. We'll look at signal coverpoints in this chapter.
- **Data coverpoints**—These capture behaviors associated with data passing through the design. For example, we might want to ensure that we tested a carry in an adder. We'll look at data coverpoints in the next chapter.

Also in the previous chapter, we saw a case where code coverage could trick us into believing we had tested a portion of our design that we had not in fact tested. We'll use a signal coverpoint in this chapter to get the real story.

Signal coverpoints are a lot like assertions, in that we describe them as transitions on one or more signals. As a result, they are usually captured by means of assertion languages such as SystemVerilog assertions (SVA).

However, SystemVerilog assertions have a very steep learning curve. It would take a completely different book to describe them properly, so we are not going to learn how to write them here. Instead, we are going to do what we did way back in Step 3, Assertions, and leverage the Open Verification Library.

The Open Verification Library (OVL) is a library of modules and components called *checkers*. In Chapter 7 we used checkers to create firewall modules and protocol monitor modules by using the assertions in the checkers.

In this chapter we use checkers to create signal coverpoints. We focus on checkers written in SystemVerilog. While there are checkers with coverage written in Verilog and VHDL, they lack the functionality to be useful. Both implementations print a list of coverpoints that have been touched.

Unfortunately, a list of coverpoints that have been touched is not as useful as a list of coverpoints that have been missed. Because the Verilog and VHDL implementations of the OVL cannot provide the list of missed coverpoints, I cannot recommend them.

That said, you can use the SystemVerilog OVL modules inside both Verilog and VHDL designs if you have a mixed-language simulator. That is the recommended solution.

27.1 Coverpoints in the OVL

Each OVL checker has an entry in the *OVL Language Reference Manual* (LRM) that describes the checker's behavior. Until now, we've used only half of that behavior, because we have turned off coverage information. Now let's dig deeper.[1]

OVL checkers are basically signal watchers. They can watch signals for one of two reasons:

1. They can act as assertions that check signals for errors.
2. They can act as coverpoints that check signals for behavior.

Each checker entry in the OVL LRM in the file `ovl_lrm.pdf` has a section called "Coverpoints." These are signal coverpoints associated with the checker. If

1. For information on installing the OVL, see Chapter 5.

the checker has no coverpoints, then this section exists but contains the word "None."

Each checker also has a section called "Covergroups," but we are not going to discuss these sections here. Their behavior becomes apparent after we discuss SystemVerilog covergroups in the following chapters.

We are going to create our signal coverpoints by finding the OVL checker that most closely matches the behavior we want to watch, and then turning on its coverage. In fact, we can leverage the checkers that we used to catch errors to provide coverage information as well.

For example, let's consider the TinyALU. In Section 26.3 on page 315, we described part of a coverage-driven test plan for this design. The test plan consisted of several coverpoints, including the following two:

- The `start` signal has been raised at least once (signal)—we watch the start signal and count the number of times it's gone high.
- The `start` signal stays high for four clock cycles (signal)—this signal coverpoint ensures that we've tested a multiple-cycle operation.

We use the OVL by knowing what all the checkers in the library do, and by choosing the one that matches our needs. If we look through the LRM, we find one checker that handles both of these coverpoints for us:

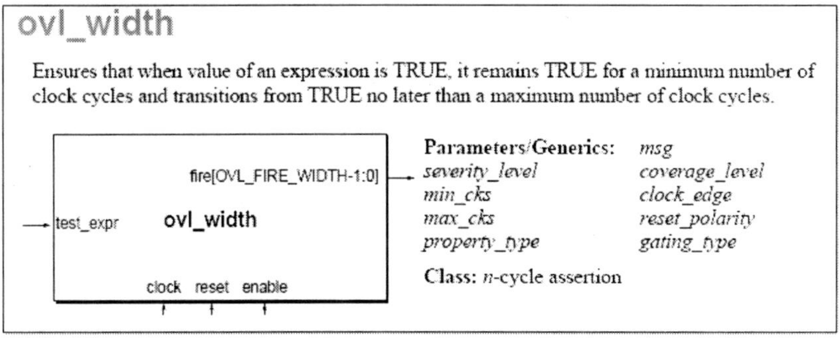

FIGURE 27-1. The `ovl_width` Checker from the OVL LRM

This checker ensures that a signal will be active for only a set range of clock cycles. We could, in fact, use this checker as part of a protocol monitor watching the TinyALU operations. We'd say that the `start` signal must be at least one clock wide and at most four clocks wide.

However, in this case we are going to use this checker only for its functional coverage capabilities. Here are the signal coverpoints in this checker:

Cover Points

`cover_test_expr_asserts`	BASIC — A check was initiated (i.e., *test_expr* was sampled TRUE).
`cover_test_expr_asserted_for_min_cks`	CORNER — The expression *test_expr* was held TRUE for exactly *min_cks* cycles (*min_cks* > 0).
`cover_test_expr_asserted_for_max_cks`	CORNER — The expression *test_expr* was held TRUE for exactly *max_cks* cycles (*max_cks* > 0).

FIGURE 27-2. The `ovl_width` Coverpoints

The checker has three coverpoints:

- `cover_test_expr_asserts`—This coverpoint makes sure that the start event was true once in the simulation.
- `cover_test_expr_asserted_for_min_cks`—This coverpoint makes sure that the signal was high for the minimum number of clocks at least once.
- `cover_test_expr_asserted_for_max_cks`—This coverpoint makes sure that the signal was high for the maximum number of clocks at least once.

In our case, these coverpoints match our needs exactly. The first coverpoint shows us that we've raised the start signal at least once. The second coverpoint tells us that we've executed a single-cycle operation. The third coverpoint tells us that we've executed a multicycle operation. If all three coverpoints have been covered, then we've tested all the operation types (though perhaps not all the operations).

However, there is a detail in Figure 27-2 that we need to discuss. The first coverpoint is a basic coverpoint, while the other two coverpoints are corner coverpoints. What do these terms mean and how do we use them?

27.1.1 OVL Coverage Levels

Different simulation jobs require different levels of "pickiness" from our functional coverage. For example, we might have been happy to know that the `start` signal in our TinyALU had gone high at least once. This is not as picky as wanting to know that it went high for the maximum and minimum number of clock cycles.

The OVL allows us to define different levels of pickiness by implementing different coverage levels. Each level looks at more-detailed aspects of coverage in a checker. There are four levels of coverage:

TABLE 27-1. OVL Coverage Levels

Level	Description	Example on a RAM
Sanity	Shows that we've activated the checker at a minimum level.	We've done a read operation.
Basic	Shows that we've done an action that tests the functionality of our object in some way.	We've written to a location and then read from it.
Corner	Simulates an action that the checker considers to be a corner case of the design.	We've accessed the maximum and minimum addresses.
Statistic	Counts interesting information about simulating our design.	Reports how many times we accessed each address.

Each checker in the OVL may have coverpoints associated with it at different levels. You tell the OVL what levels of coverage you want, and all the checkers that implement those levels of coverage create coverpoints.

We control coverage levels by using parameters and macros. Each OVL checker has a parameter called `coverage_level`. We can set the coverage level on a checker-by-checker basis by explicitly passing each checker its coverage level as a parameter.

The levels themselves are defined as macros. We pass the macros to the checker through the `coverage_level` parameter. There are six macros that control the coverage levels:

TABLE 27-2. OVL Coverage-Level Macros

Coverage Level	Macro Name	Value
None	`OVL_COVER_NONE`	`4'b0000`
Sanity	`OVL_COVER_SANITY`	`4'b0001`
Basic	`OVL_COVER_BASIC`	`4'b0010`
Corner	`OVL_COVER_CORNER`	`4'b0100`
Statistic	`OVL_COVER_STATISTIC`	`4'b1000`
All	`OVL_COVER_ALL`	`4'b1111`

Each coverage value represents one bit in a four-bit number that holds the coverage level. These coverage values can be combined by ORing the bits together. For example, let's create a value for sanity checking and basic checking:

```
`OVL_COVER_SANITY|`OVL_COVER_BASIC
```

This creates the value `4'b0011`, which will cause the checker to use both sanity coverage and basic coverage.

27.2 Instantiating OVL Checkers for Coverage

We first looked at OVL checkers as a way of implementing assertions that would make it easier to debug our designs. Now we are looking at them as ways of implementing signal coverpoints to make sure that we've tested our design fully.

In some cases, we may want a checker to do double duty. We'll want it to check that our signals are behaving properly, and we'll want it to check that we tested the signals.

In other cases we may want to turn off the checking or assertion capability in a given checker. We can do that on a checker-by-checker basis by using parameters.

There are two parameters that control what a checker does regarding assertions and coverage:

- `property_type`—By default this parameter is set to the OVL macro `OVL_ASSERT`. This macro tells the checker to implement assertions. We can also set this parameter to `OVL_IGNORE`. This turns off assertions.
- `coverage_level`—This parameter can take any of the values from Table 27-2 on page 324. This includes `OVL_COVER_NONE`, which turns off coverage.

We use a strategy for instantiating functional coverage checkers that is similar to the strategy we used for assertions. We'll create a separate functional coverage module with inputs that contain all the relevant signals. Then we can instantiate that module into either a Verilog or VHDL top level.

Here is the functional coverage module that checks our start signal:

```
16  `include "std_ovl_defines.h"
17
18  module alu_functional_coverage
19    ( input bit clk,
20      input bit reset_n,
21      input bit done,
22      input bit start,
23      input logic [2:0] op,
24      input logic [7:0] A,
25      input logic [7:0] B,
26      input logic [15:0] result);
27
28      wire [2:0] fire;
29
30      ovl_width #(.min_cks(1), .max_cks(4),
31                  .property_type(`OVL_IGNORE),
32                  .coverage_level(`OVL_COVER_BASIC | `OVL_COVER_CORNER))
33
34         multi_clock(.clock(clk), .reset(reset_n), .enable (reset_n),
35                     .fire(fire),
36                     .test_expr(start));
37  endmodule
```

FIGURE 27-3. Functional Coverage Module

- **Line 16**—We need access to the macro definitions that come with the OVL, so we include this file.
- **Line 18**—We are creating a module to hold all our functional coverage.
- **Lines 19–26**—Input all the signals from the top level.
- **Line 28**—These three bits are a placeholder for the `fire` bus.
- **Line 30**—Instantiate `ovl_width` and set the `min_cks` and `max_cks` parameters.
- **Line 31**—Turn off assertions in this checker.

- **Line 32**—We need basic and corner coverage to create all three signal coverpoints.
- **Line 36**—We are testing the `start` signal.

This pattern works for all signal coverpoints that we instantiate with the OVL. We turn off the assertions, turn on the levels of coverage we want, compile, and simulate.

27.2.1 Compiling OVL with Functional Coverage

When we compile the OVL, we use macros to turn on the assertions and the coverage. We define the macros at compile time to turn on certain features. Here are the Questa compilation options that we used for our functional coverage point:

```
8  +libext+.vlib
9  +libext+.v
10 +incdir+../../std_ovl
11 +define+OVL_ASSERT_ON
12 +define+OVL_COVER_ON
13 +define+OVL_SVA
14 +define+OVL_MAX_REPORT_COVER_POINT=0
15 -y ../../std_ovl
```

FIGURE 27-4. Questa Compilation Options for OVL Functional Coverage

- **Lines 8–11**—These lines are the same as those we used for assertions.
- **Line 12**—This is new. We now want the checkers to implement coverage, so we define the `OVL_COVER_ON` macro.
- **Line 13**—We want SVA so we define the `OVL_SVA` macro.
- **Line 14**—Because we are using SVA, we don't want the checker to print its coverage information to the screen. This saves time.
- **Line 15**—The checker modules are defined in the `../../std_ovl` directory.

27.3 Simulating with OVL Functional Coverage

When we simulate a design with functional coverage turned on, we add SystemVerilog coverage directives to our simulation. Different simulators handle these coverage directives differently. We're going to examine how they look in the Questa GUI.

We run the same test that we ran at the beginning of Figure 26 on page 311. Recall that this test simulated all the operations except the multiply. We noticed that, even though the multiply had not been simulated, its block had 100% code coverage. We wanted to find a better way of measuring coverage, to avoid being tricked by situations like this.

We see in the figure below how signal coverpoints help solve this problem:

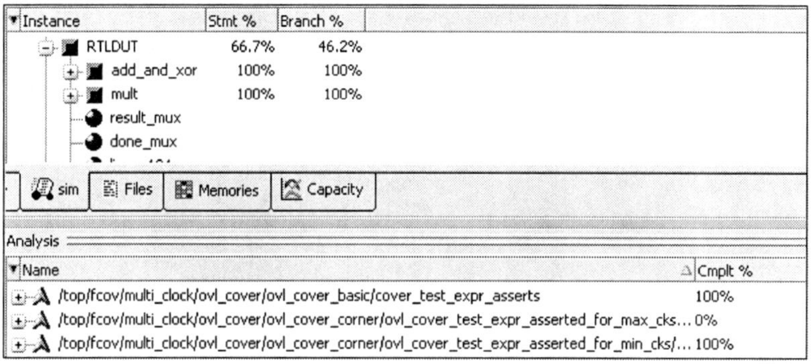

FIGURE 27-5. Functional Coverage in the Questa GUI

We see that the `add_and_xor` and `mult` blocks still have 100% coverage; however, now we see a clear sign that we haven't simulated everything. The `ovl_cover_test_expr_asserted_for_max_cks` coverage directive is at 0%. We never tested the case where the start signal was high for multiple clocks.

27.4 Summary

In this chapter we learned how to create signal coverpoints with the Open Verification Library (OVL).

We saw that each checker in the OVL has a section called "Coverpoints," which describes the signal coverpoints implemented by that checker. We can implement coverpoint directives in a coverage-driven test plan by finding the OVL checker that most closely matches our needs, and then instantiating it.

We also learned that each instance of an OVL checker can control whether that instance implements assertions, coverage, or both. We control the checkers through parameters that control their behavior.

We saw that it was best for us to use the SystemVerilog assertion checkers for functional coverage. This is because SystemVerilog assertions allow simulators to deliver reports that show us which functional coverage points we missed. The Verilog and VHDL versions of coverage cannot do this. Finally, we saw the macros we need to define when we compile our checkers to turn on coverage and control its output.

This chapter gave us an introduction to OVL checkers and their uses in functional coverage. This is a large topic, and there are many more examples and articles available on test benches at www.fpgasimulation.com.

In this chapter, we ignored the section of the OVL LRM titled "Covergroups." This is because we are going to learn how to write our own covergroups in the following chapters.

28 Introduction to Covergroups

IN THIS CHAPTER:

- *Understanding Covergroups*
- *Adding a Covergroup to the* `tester`
- *Simulating with Covergroups*

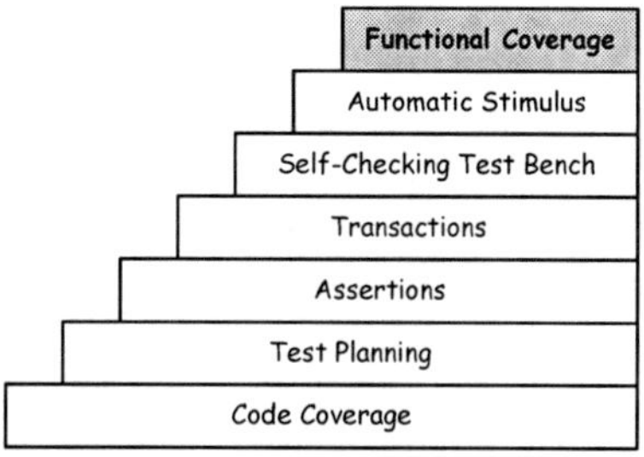

In our previous two chapters, we looked at the failings of simple code coverage and saw that we need to augment code coverage with functional coverage. We also saw that functional coverage can be part of a new kind of test plan called a *coverage-driven test plan.*

In a coverage-driven test plan, you describe all the functions that need to be tested in terms of coverpoints. Then you use the simulator to make sure that all the coverpoints have been tested. If you have a self-checking test bench, and you've tested all the coverpoints, then you've tested your design.

There are two kinds of coverpoints:

- **Signal coverpoints**—These tell us that we've exercised the signals that we want to see exercised in our design. We implement these coverpoints with OVL checkers.
- **Data coverpoints**—These tell us that we've passed through our design the data that tests it fully. We implement these coverpoints with SystemVerilog covergroups.

In this chapter, we introduce SystemVerilog *covergroups*. We learn how to use covergroups to implement data coverpoints, and how to create robust coverage models for our design.

We can use covergroups to demonstrate to ourselves (and to others) that we've tested all the behaviors and data that we expect to see in our design.

28.1 Understanding Covergroups

While covergroups share similarities with classes and objects, they have a new kind of behavior insofar as they do nothing but count the number of times we've seen certain pieces of data or combinations of data. Covergroups consist of coverpoints that connect to variables and expressions, as well as bins that count the number of times those expressions take on a value. We will discuss them in detail below.

We define covergroups inside of modules and use them to monitor the data that passes through the module variables. This gives us the opportunity to place them in several places:

- We can place covergroups inside a transaction generator, to make sure that we've generated all the interesting transactions we want.
- We can place covergroups inside a driver, to make sure that we've passed the proper data onto the RTL signals.
- We can place covergroups inside a monitor, to make sure that we've seen all the necessary data come out of an RTL DUT.

In our example, we place a covergroup inside the `alu_tester` module to make sure that we've given the TinyALU a run for its money. We'll use this example to learn about covergroups and to build a robust coverage-driven test bench for the TinyALU.

28.1.1 Covergroups, Coverpoints, and Bins

Before we can use covergroups, we need to understand the terms *covergroup*, *coverpoint*, and *bin*.

A *covergroup* is a SystemVerilog construct that looks a lot like an object. To use a covergroup, we define it in the module and then declare a variable to hold a handle to it. Finally, we create a new instance of the covergroup with the new() method.

Covergroups watch the variables inside a module. We tell them which variables to watch by creating *coverpoints*. We define coverpoints in terms of expressions built from the module variables. The expression can be as simple as naming the variable, or it can be a combination of variables.

Once we've told the covergroup what variable we're watching, we need to define the values of interest in that variable. We do that by creating *bins*, which define data in the variable. When the variable contains data that matches a bin definition, the covergroup increments the bin. For example, we might make a bin that watches for whether the `A` leg of the TinyALU is all zeros.

Once we've defined the covergroups, coverpoints, and bins, then our goal is to make sure that every bin gets incremented at least once. If we've touched every bin in a covergroup, then we have achieved 100% coverage for that covergroup.

We'll see exactly how to do these things by walking through our TinyALU example.

28.2 Adding a Covergroup to the `tester`

We're going to learn about covergroups by adding one to the TinyALU test bench, putting it into the `alu_tester` module. The covergroup ensures that we've tested all our operations.

Our first step is to create a coverpoint description for the TinyALU coverage plan. Here's the coverpoint description we'll implement in this section:

Make sure we've simulated every operation.

Now we're ready to create a covergroup. There are three steps to using a covergroup:

1. Define the covergroup, including its coverpoints and bins.
2. Instantiate a new covergroup and put it into a variable.
3. Tell the covergroup to sample the data in the module and increment the bins.

Once we've implemented these three steps, we run the simulation and make sure all the bins are covered.

28.2.1 Defining the Covergroup

We define the covergroup in a module in the same way that we create any other user-defined type. The definition goes at the top of the module. Then we'll declare a variable of that type, and finally we'll instantiate a new covergroup into the variable.

We define a covergroup with the following syntax:[1]

```
covergroup <name>;
   <coverpoint_name> : coverpoint <expression> {
      bins <bin_definition> }
endgroup
```

The `bins` definition is optional. As we'll see below SystemVerilog can automatically create bins.

Here is a covergroup in our `alu_tester` module:

```
24
25  module alu_tester(ref tlm_fifo #(alu_operation) op_f);
26
27  test_request req;
28  operation_t   op;
29
30  covergroup op_cov;
31     coverpoint op;
32  endgroup
33
34  op_cov oc;
35
```

FIGURE 28-1. Covergroup Definition for TinyALU

- **Line 25**—This is the module in our transaction-level test bench that creates new transactions and puts them into the `op_f` FIFO.
- **Line 27**—The transactions go into a variable called `req`.
- **Line 28**—Covergroups cannot see data in objects, so we copy the data from the object to a variable such as `op` so that the covergroup can see it.
- **Line 30**—We define a covergroup called `op_cov`.
- **Line 31**—We create a coverpoint inside of `op_cov` that watches the `op` variable.

1. This is a subset of the full syntax, but we'll use it for clarity.

- **Line 32**—We end the group. Notice that we did not define the bins. This is because `operation_t` is an enumerated type. This allows SystemVerilog to create the bins for us. It automatically creates one bin per operation.
- **Line 34**—We declare the variable `oc` of type `op_cov`. This variable holds our covergroup.

The next step is to instantiate a copy of the covergroup.

28.2.2 Creating a New Covergroup

Covergroups are like objects in that we need to call a constructor and create them before we can use them. Like objects, the constructor is a method called `new()`, and we assign its output to a variable that holds a handle to our type.

Here's how we instantiate a covergroup in the `alu_tester` module:

```
34 op_cov oc;
35
36 string s, m;
37    task run;
38       oc = new();
```

FIGURE 28-2. Instantiating an `op_cov` Covergroup

- **Line 34**—Declare a variable to hold our covergroup.
- **Line 38**—Call `new()` to put a new covergroup into that variable.

Now that we have a variable holding the covergroup, we can tell the covergroup when to sample the data in the module.

28.2.3 Sampling Data with a Covergroup

Covergroups sample the data in a module. We link the covergroup to the variables by using the variable names in coverpoint expressions. For example, we connect the `op` variable in our module to our covergroup in Figure 28-1 on page 332, line 31. We did it by creating a coverpoint with the `op` variable in the coverpoint expression.

We've put an instance of that covergroup into the `oc` variable. Now we need to tell `oc` when to sample. We do that with the `sample()` method that exists in all covergroups. Here's an example with the `alu_tester` module:

```
40        req = new();
41        repeat (1000) begin
42           assert(req.randomize());
43           op = req.op;
44           oc.sample();
```

FIGURE 28-3. Sampling with a Covergroup

- **Line 40**—Create a new transaction object.
- **Line 41**—We put 1000 randomized objects into the test bench.
- **Line 42**—Randomize the transaction.
- **Line 43**—Copy the operation from the transaction into the `op` variable.
- **Line 44**—Call `sample()` to check the `op` variable and increment the right bins. SystemVerilog automatically creates one bin per operation.

This pattern is common whenever we use a covergroup. We put the covergroup into a module, create coverpoint expressions from module variables, and then we call `sample()` to count up how many times we've met our coverage goals.

28.3 Simulating with Covergroups

When we first started looking at functional coverage, we demonstrated a problem with code coverage. We had tested the TinyALU with a test bench that was guaranteed not to test the multiply operation, yet the code coverage for the multiply block showed 100% coverage.

We addressed this problem somewhat with a signal coverpoint implemented in the OVL. That coverpoint showed us that we had never tested a multicycle operation. Because there was only one multicycle operation, we knew that we had not tested the multiply.

However, there are two problems with this approach. The first is that this was an indirect way to learn that we hadn't tested the multiply. The second problem was that we had no way of knowing directly whether we had tested all the single-cycle operations. We could infer this from the 100% code coverage, but we might want something more direct.

We've now addressed that problem by adding a covergroup to the `alu_tester` module. The covergroup implements this statement in our test plan:

Make sure we've simulated every operation.

Now we can simulate the design and look at our covergroup results to see whether we've tested every operation.When we run the TinyALU test bench in Questa,[2] we get the following covergroup report in the GUI:

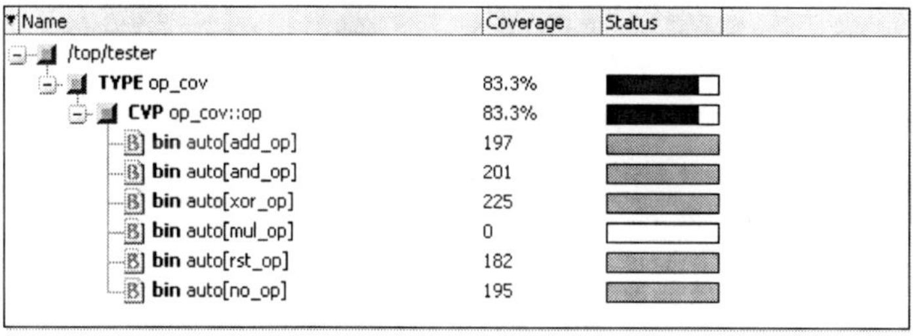
Name
Coverage
Status
/top/tester
TYPE op_cov
83.3%
CVP op_cov::op
83.3%
bin auto[add_op]
197
bin auto[and_op]
201
bin auto[xor_op]
225
bin auto[mul_op]
0
bin auto[rst_op]
182
bin auto[no_op]
195

FIGURE 28-4. Questa Covergroup Report for TinyALU

This report shows us that we did not meet our goal from the coverage test plan. We never simulated the multiply operation.

Each time we called `sample()`, the covergroup looked at the `op` variable and incremented the bin that matched its value.

The covergroup gives us a complete picture of what's going on in our test bench. When we combine data coverage with signal and code coverage, we can completely analyze what we've tested.

2. Other simulators have different ways of displaying the information from the covergroups.

We saw above that we didn't test multiply. This is in agreement with our signal-level coverage, which shows that we never simulated a multicycle operation:

Name	Cmplt %
/top/fcov/multi_clock/ovl_cover/ovl_cover_corner/ovl_cover_test_expr_asserted_for_max_cks...	0%
/top/fcov/multi_clock/ovl_cover/ovl_cover_basic/cover_test_expr_asserts	100%
/top/fcov/multi_clock/ovl_cover/ovl_cover_corner/ovl_cover_test_expr_asserted_for_min_cks/...	100%

FIGURE 28-5. TinyALU Signal Coverpoints

The missing data and signal coverage explain the missing code coverage below:

Instance	Stmt %	Branch %	Condition %	Expression %
top				
RTLDUT	66.7%	66.7%		
add_and_xor	100%	100%	80%	
mult	100%	100%		33.3%

FIGURE 28-6. TinyALU Code Coverage

We never fully tested the top level because we never ran a multiply.

28.4 Summary

In this chapter we learned how to create a simple covergroup. We saw that we can define covergroups inside of a module, and that the covergroup monitors the variables inside the module.

We saw that there are three steps to using a covergroup:

1. Define the covergroup and declare a variable to hold it.
2. Instantiate the covergroup into the variable
3. Run the simulation and call `sample()` to increment the bins.

We learned that covergroups consist of coverpoints, which are defined with expressions that use module variables. Each coverpoint has bins, which look for expression values and increment when those values appear.

In our example, we created a simple coverpoint to watch the operations that went into the TinyALU. Because the variable had an enumerated type, SystemVerilog was able to create a bin for each possible value, and we got a report showing that we had never simulated the multiply function. We saw that a covergroup is 100% covered only when all its bins have been incremented.

Now that we have the basics of covergroups under our belt, we can use covergroups to create complex coverpoints. In our next chapters, we'll create additional coverpoints for the TinyALU. For example, does the TinyALU work if we run the same operation twice in a row?

29 Defining Data Bins

IN THIS CHAPTER:

- *Using Automatic Bins in Numeric Coverpoints*
- *Defining Data Bins*
- *Using Bins with Ranges of Values*

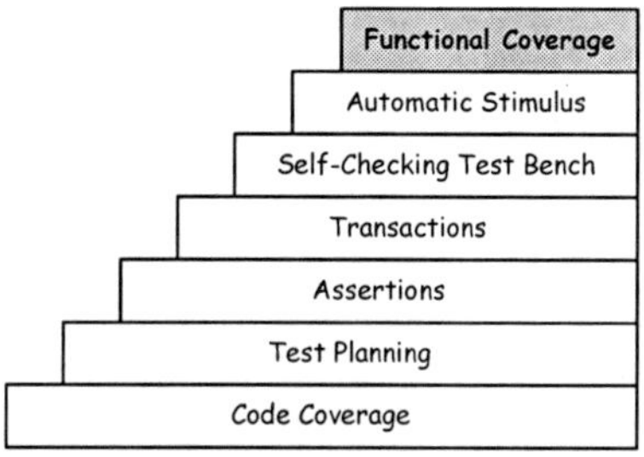

In the preceding chapter, we learned how to add a covergroup to a module to create data coverpoints. We used the covergroup to check that we had created transactions containing all the possible operations, and we learned that we were not creating a multiply transaction.

That first example of a covergroup relied on SystemVerilog to create our bins. SystemVerilog knew that `op` was an enumerated type, so it created one bin for each value in the enumerated type. On the other hand, letting SystemVerilog do our work can take us only so far. It's time to create our own bins.

Let's add another line to our coverage-driven test plan. We know that ALUs can have trouble at the corners of their data range, so let's add the following coverage point:

Test the case where the A or B legs are 8'h0 and 8'hF.

This requires that we add another coverpoint to our `alu_tester` covergroup.

29.1 Using Automatic Bins in Numeric Coverpoints

We are trying to create a covergroup that tells us whether or not we've tested the TinyALU with `'h00` or `'hFF` on one or both legs. In the previous chapter, we simply created a coverpoint with the same name as the variable that interested us, and that solved the problem. Let's try that again. We'll add an `A` and `B` variable to the `alu_tester` module and then add two new coverpoints to our covergroup:

```
31  covergroup op_cov;
32
33     coverpoint op;
34     coverpoint A;
35     coverpoint B;
36
37     option.auto_bin_max = 4;
38
39  endgroup
```

FIGURE 29-1. Coverpoints for `A` and `B` in the TinyALU

- **Line 34**—This is a new coverpoint that handles the `A` variable. `A` is an eight-bit number.
- **Line 35**—This new coverpoint handles the `B` variable. This is an eight-bit number.
- **Line 37**—This limits the number of bins to four.

This covergroup now makes automatic bins to cover `A` and `B` as well as `op`. We've asked SystemVerilog to create bins for these variables automatically, and this is why we have the option on line 37.

When we ask SystemVerilog to create bins automatically for a numeric value, it divides the range of that value into 64 bins. In our case, `A` and `B` had 256 possible values, so we would have obtained 64 bins with four values each. This seems excessive, so we asked SystemVerilog to create only four bins by defining a covergroup option.[1]

1. There are many options available for covergroups, and they can be quite arcane and detailed. There are more discussions of these options on www.fpgasimulation.com.

We are going to run our simulation with 100 randomized transactions. Here is the output showing the automatically created bins in the Questa GUI:

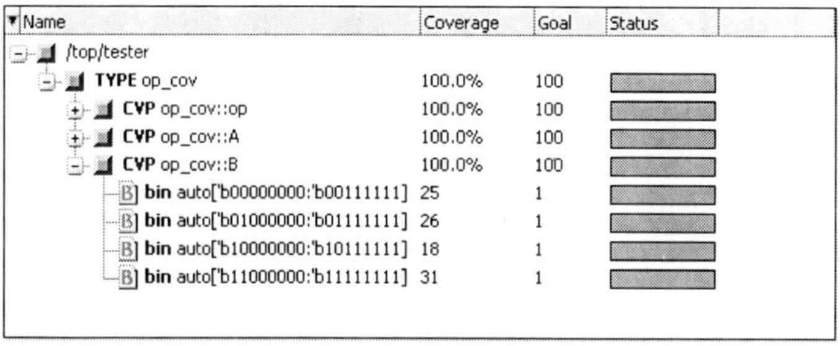

FIGURE 29-2. Automatically Created Bins for Data

Well, this is a mildly interesting piece of data. It shows us that we tested all our operations, and it shows us that we've tested numbers in the four quartiles for the `A` and `B` legs. Unfortunately, it doesn't meet the basic objective:

Test the case where the A or B legs are 8'h0 and 8'hF.

It's time to define bins explicitly to solve the problem.

29.2 Defining Data Bins

As seen in the previous chapters, we define bins in coverpoints with the `bins` keyword. The definition looks like this:

```
<coverpoint_name> : coverpoint <expression> {
   bins <bin_name> = {<list of values>};
}
```

We use the coverpoint the same way that we did for default bins. It accepts an expression made of variables in the module. The `bins` declaration is new. We add bins by putting a pair of braces, `{ }`, after the coverpoint and defining the bins within the braces.

Each bin starts with the keyword `bins`. Then there is the name of the bin and another pair of braces that contain a list of numbers. Finally, there is a semicolon after each bin declaration. This leads to the same strange "`};}`" syntax that we saw in the constraints chapter. *C'est la vie.*

Let's create an example of a coverpoint with explicit bins. We want to make sure we saw a case where there were `'hFF` or `'h00` on `A` or `B`. There are two ways to do this. We'll use one method for `A` and the other for `B`:

```
31  covergroup op_cov;
32
33     coverpoint op;
34
35     A_00_FF: coverpoint A {
36        bins zeros = { 0 };
37        bins ones  = { 'hFF };}
38
39     B_00_FF: coverpoint B {
40        bins zero_or_ff = {0, 'hFF};}
41
42  endgroup
```

FIGURE 29-3. Creating Single Bins

- **Line 35**—Notice that we've named our coverpoint now. We've also added the "`{`" to say that we are defining bins for this coverpoint.
- **Line 36**—Define a bin that increments if `A` is `'h00`.
- **Line 37**—Define a bin that increments if `A` is `'hFF`.
- **Line 39**—We have named our `B` covergroup.
- **Line 40**—Define a single bin that increments if `B` is `'h00` or `'hFF`.

This example demonstrates the simplest way to define a bin. We make a list of all the values that are interesting to the bin. The bin increments on a sample if the expression on the `coverpoint` line matches a value in the list.

We saw two strategies at use in this example. One of them creates a bin for the case where A is `'h00`, and a separate bin for the case where A is `'hFF`. The other strategy is to have one bin for both. The choice between the strategies depends upon what is important to you.

Let's run this simulation and look at the results:

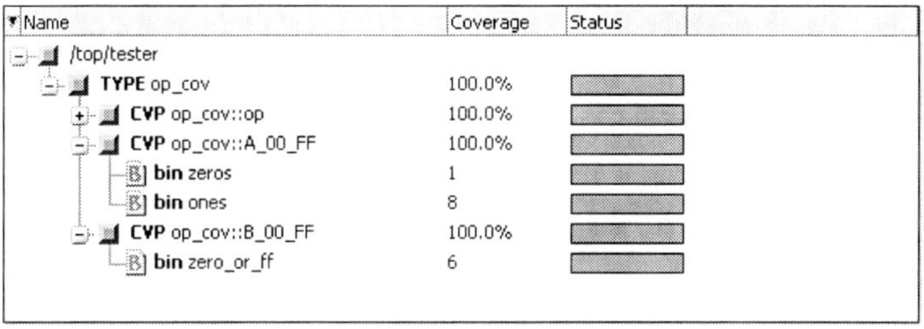

FIGURE 29-4. Looking for `'h00` and `'hFF`

We've run 1000 transactions through the TinyALU with no constraints. We see that we've achieved all our coverage goals. We've tested every type of operation and we got `'h00` and `'hFF` on our inputs.

However, notice the difference in information that we have for `A` and `B`. With `A` we know that we had `'h00` only once out of a thousand transactions, but we had `'hFF` eight times. With `B` we know that we had `'h00` or `'hFF` on the input six times collectively, but we don't know the breakdown.

29.3 Bins with Ranges of Values

In our last example, we created bins by providing a list of numbers that were interesting to us: `'h00` or `'hFF`. In addition, there are many cases where we are more interested in a range of values rather than single values. Covergroups allow us to define bins using ranges for that reason.

Let's use bins to add another rule to our coverage list:

Simulate at least one single-cycle operation
and one multicycle operation.

This is a superset of our rule that we need to simulate all the operations. In this case we just want to know that we got each kind of operation. We can create a bin to handle this by using ranges.

Before we create the new bin, let's take a look back at our definition of the operation for TinyALU. These were defined by means of an enumerated type called `operation_t`. Here is that definition:

```
typedef enum {add_op, and_op, xor_op, mul_op, rst_op, no_op} operation_t;
```

FIGURE 29-5. Operation Enumerated Type

We're looking at this definition to note the order in which the operations were defined. Notice that all the single-cycle operations are together, then the multicycle operation (`mul_op`), and finally the miscellaneous operations. We can use this default ordering to create our covergroup:

```
31 covergroup op_cov;
32
33    coverpoint op {
34       bins single_cycle = {[add_op : xor_op], rst_op,no_op};
35       bins multi_cycle = {mul_op};}
36
```

FIGURE 29-6. Using Ranges in Bin Declarations

- **Line 34**—We've created a bin by using a combination of a range and single values. We needed to skip over the `mul_op` in the enumerated type, so we used a range for the first three operations and then single entries for the last two.
- **Line 35**—This is a single-value bin that captures the multiply operation.

In this example, all the values in a range combine to create a single bin. As a result, we have one bin that increments for all single-cycle operations and another bin that increments for the multicycle operation.

However, when we run eight transactions through the test bench, we see the following limitation of this approach:

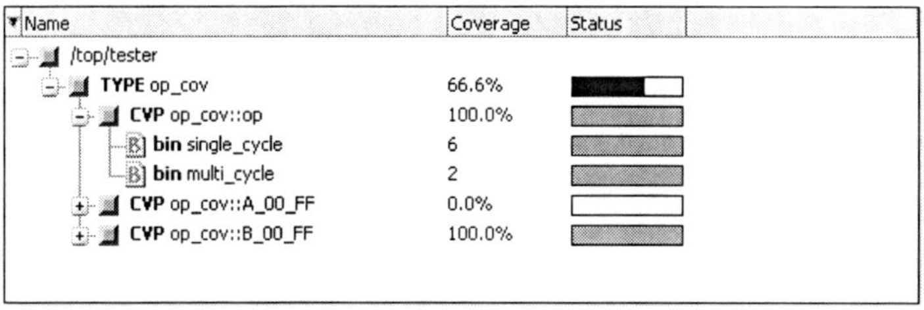

FIGURE 29-7. Range-Based Bins in Action

We've run eight transactions through the TinyALU, and we've achieved the goal of testing at least one single-cycle operation and one multicycle operation. Unfortunately, there is no way to tell whether we've tested all our operations; this is because when we defined bins explicitly, we lost the automatic bins. Fortunately, SystemVerilog allows us to capture that information as well.

29.3.1 Generating Automatic Bins from Ranges

We have 100% coverage with only eight transactions: two `multi_cycle` transactions and six `single_cycle` transactions. This example shows us how the coverage metric works. It counts the number of bins that have reached their goal and divides by the total number of bins. We've touched each bin once, so we've achieved 100% coverage.

But have we really achieved this lofty coverage? We know for a fact that there are five single-cycle operations available (`and_op`, `add_op`, `xor_op`, `rst_op`, `no_op`). Did we test all of them? Wouldn't it be more comforting to know that we hit all our operations?

We can modify our bins to make this happen. Up until now we've created one bin for each value or set of values. It's time to make the `bins` keyword live up to its name and create multiple bins. We can do that by inserting square brackets, `[]`, into our `bin` declaration:

```
bins onepervalue [] = {<list of values>};
```

The square brackets are the magic here. The empty square brackets tell SystemVerilog to create one bin per value in the list.

We can set the number of bins ourselves, as follows:

```
bins n_bins [n] = {<list of values>};
```

This line creates *n* bins. If there are more values than `n`, then the values are divided evenly across the bins. If the number of values is not divisible by the number of bins, then the last bin gets the remainder.

The list of values can incorporate ranges, and if there are multiple ranges they can overlap. SystemVerilog expands the ranges into a list. So, for example, this definition:

```
bins threebins[3] = {[1:2],[2:6]};
```

is the same as this definition:

```
bins threebins[3] = {1,2,2,3,4,5,6}
```

Both create three bins that respond to the values `<1,2>`, `<2,3>`,`<4,5,6>`. If we had not specified the `[3]`, then SystemVerilog would have made seven bins.

We'll modify the coverpoint so that some of our bins use the square brackets, `[]`. This gives us not only a measure of whether we've tested single-cycle and multi-cycle operations, but also a measure of whether we've tested all the individual operations. We also use the square bracket operator to solve our problem with the `B` leg of our `'h00` or `'hFF` coverage.

We can now let SystemVerilog create a bin for each of these scenarios:

```
31 covergroup op_cov;
32
33    coverpoint op {
34       bins single_cycle[] = {[add_op : xor_op], rst_op,no_op};
35       bins multi_cycle = {mul_op};}
36
37    A_00_FF: coverpoint A {
38       bins zeros = { 0 };
39       bins ones  = { 'hFF };
40    }
41
42    B_00_FF: coverpoint B {
43       bins zero_or_ff[] = {0, 'hFF};
44    }
45
46 endgroup
```

FIGURE 29-8. Multiple Bins Per `bins` Line

- **Line 34**—Create a bin for each of the single-cycle operations.
- **Line 43**—Create a bin for the `'h00` and `'hFF` case on the `B` leg.

Now we'll get a bin for each operation:

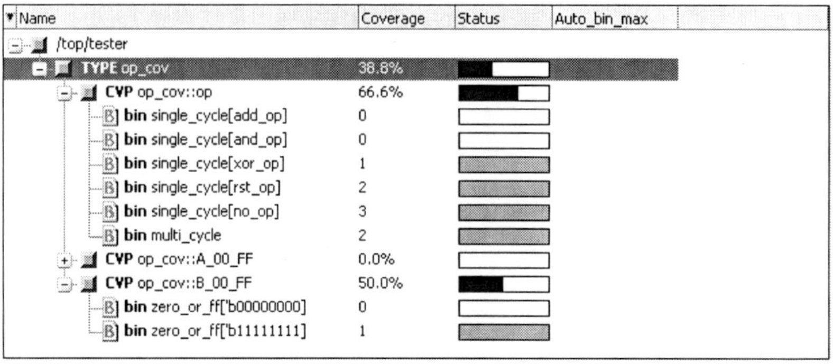

FIGURE 29-9. Coverage with Multiple Bins

Well, this is not nearly as cheery a picture as that in Figure 29-7 on page 345. Our eight transactions leaned heavily towards `no_op` and `rst_op`. We did test the multiply. We can also see that while we hit the `zero_or_ff` bin on `B`, we did it only once, and only for `'hFF`.

29.4 Summary

In this chapter, we continued our investigation of data coverpoints by using covergroups. We learned how to create more-detailed coverage models by defining our own bins explicitly in coverpoints.

We saw that we could define bins by specifying the values associated with them, and that we could lump multiple values together into one bin. We also saw that we could use ranges to define bins.

While it is useful to be able to lump multiple values into a single bin, it is also useful to take advantage of SystemVerilog's ability to create bins automatically along with lists of values. We saw that we could create a coverage model that would tell us whether we had tested single-cycle and multicycle operations, and that could check on the number of times we simulated each operation.

However, as valuable as this coverage is, it would also be valuable to be able to cover the transitions from one data value to a different data value. For example, how do we know that the multiply operation responds properly if it is executed after a reset? Is data left in the pipeline? In our next chapter, we'll learn how to add transitions to our coverage models.

30 Transition Coverage

IN THIS CHAPTER:

- *Single-Value Transitions*
- *Multiple-Value Transitions*
- *Repetition Coverage*

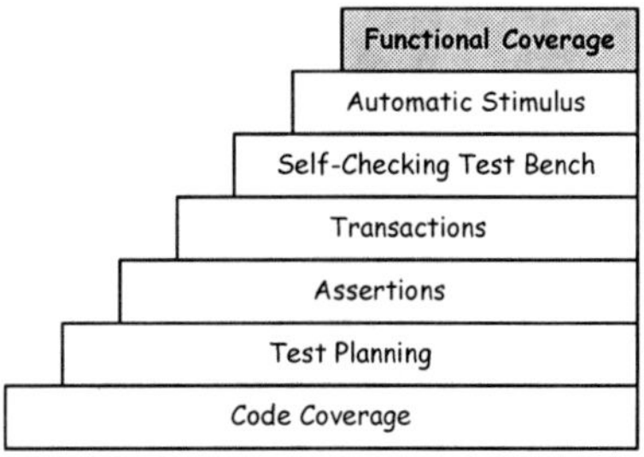

Functional coverage turns the usual approach to test writing on its head. With normal test writing, you think of something that needs to be tested, and then you write a test to do the job. That is a slow way to test large designs.

Coverage-based test plans have a different strategy: you decide what needs to be tested and then write a coverpoint to make sure it has been tested. Then you let the automatic stimulus in your test bench do the actual testing. That way, you don't need to write a separate test for every piece of functionality. You need to write only the functional coverage.

So far, we've been able to handle coverpoints that need to see just a certain piece of data. For example, we wrote a TinyALU test bench that checks to see whether we've simulated every operation. However, this is just a first-order test of the design. Let's consider another question.

The TinyALU has two blocks. One, the single-cycle block, returns the result in the same clock cycle as the operation request. The other, the multiply block, returns the result in three clock cycles. The multiply block is a pipelined multiplier. This allows the device to run with a higher clock speed, even though we need to wait a few clocks for a product.

Pipelined operations raise new verification questions. Specifically, we need to check that the pipeline is being cleared properly between operations and on resets. This realization creates four new coverpoints in our test plan:

Have we tested all the operations before and after a reset?
Have we tested the same operation twice in a row?
Have we run a single-cycle operation before a multicycle operation?
Have we run a multicycle operation before a single-cycle operation?

We cannot test for these coverpoints with the SystemVerilog features we've learned, because all the bins we've created look for a single value. We need to create bins that look for transitions between values.

30.1 Single-Value Transitions

Transition coverage allows us to create bins that increment when a variable transitions from one value to another. The coverpoint names the variable to watch, and then we create a bin that increments on the correct transition. Here's the basic syntax for creating a transition bin:

```
bins bin_name (<value1> => <value2>);
```

This bin looks for `value1` and then, on the next sample, it looks for `value2`. If it gets `value2` it increments the bin. (Notice that we use regular parentheses here, not curly braces.)

Here's an example. Let's add bins to our `op` coverpoint to check that we see a `mul_op` before and after a `rst_op`:

```
coverpoint op {
   bins single_cycle[] = {[add_op : xor_op], rst_op,no_op};
   bins multi_cycle = {mul_op};

   bins mult_rst = (mul_op => rst_op);
   bins rst_mult = (rst_op => mul_op);

   }
```

FIGURE 30-1. Transition Bins for Watching Multiply and Reset Operations

- **Line 37**—This is the same `op` coverpoint that we used before, but now we're adding a bin to look for transitions between `mul_op` and `rst_op`.
- **Line 38**—Add a bin to look for transitions between `rst_op` and `mul_op`.

This code creates two new bins that watch the `op` variable. One bin increments if `op` transitions from multiply to reset. The other increments if `op` transitions from reset to multiply. This covergroup gives us the following report after 100 transactions:

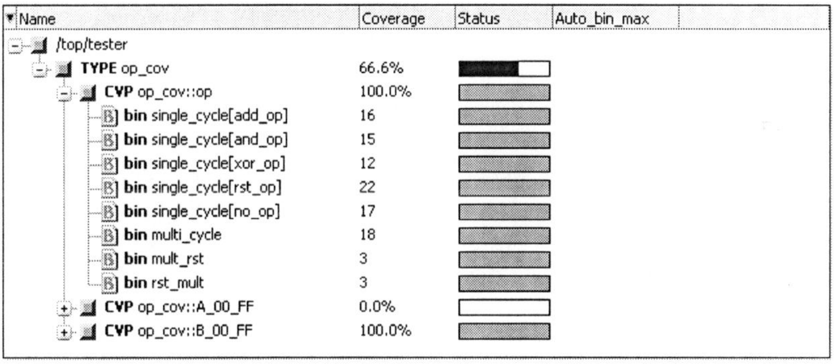
Name
Coverage
Status
Auto_bin_max
/top/tester
TYPE op_cov
66.6%
CVP op_cov::op
100.0%
bin single_cycle[add_op]
16
bin single_cycle[and_op]
15
bin single_cycle[xor_op]
12
bin single_cycle[rst_op]
22
bin single_cycle[no_op]
17
bin multi_cycle
18
bin mult_rst
3
bin rst_mult
3
CVP op_cov::A_00_FF
0.0%
CVP op_cov::B_00_FF
100.0%

FIGURE 30-2. Result of Transition Bin

Sure enough, in 100 transactions we tested both transitions three times. The nice thing about functional coverage with automatic stimulus is that I don't need to create a test to capture this functionality explicitly. I just need to make sure that the transitions between the reset and multiply operations occurred. The test bench is self-checking, and because it didn't generate an error, I know that my reset and multiply worked properly.

Of course, this doesn't fully meet our functional coverage requirement. We said that we wanted to test all operations before and after a reset. We could add more bins, one for each operation, but there is an easier way to test transitions on multiple values.

30.2 Multiple-Value Transitions

Creating transitions with single values on either side of them got us started with transition coverage, but with a large number of possible operations this would lead to a ridiculous amount of typing. For example, consider the second coverage objective from our coverage-driven test plan:

Have we tested all the operations before and after a reset?

There are six operations in the TinyALU. To test all of them before and after a reset would require that we explicate create twelve bins. Fortunately, SystemVerilog allows us to specify lists of transitions, after which it does all the work. Here is the syntax for creating a transition with multiple values:

```
bins bin_name[] = (<value list> => <value list>);
```

This is an extension of the single-value example. Now, instead of using a single value for the transition, we can provide a list of values on both sides of the transition. SystemVerilog expands the list into all its combinations. (Notice also that we can create multiple bins by using the square brackets.)

SystemVerilog takes our transition lists and expands them out in a way that is similar to multiplying equations in algebra. It creates a single-value transition from every value on the left side to every value on the right side.

For example, SystemVerilog expands the following multiple-value transition:

```
1,2 => 3,4,5,7
```

into the following single-value transitions:

```
1=>3, 2=>3, 1=>4, 2=>4, 1=>5, 2=>5, 1=>7, 2=>7
```

You can also use ranges for your list just as you can when making bins. So we could have written the above like this:

```
1,2 => [3:5],7
```

If we wanted a separate bin for each transition, our `bins` line should look like this:

```
bins manybins[] = (1,2 => 3,4,5,7);
```

The square brackets on this line created many bins. If we don't use the square brackets, we'll get one big bin that increments on any of the transitions. Let's modify our coverage model to make sure that we've tested all our operations before and after a reset:

```
33       coverpoint op {
34          bins single_cycle[] = {[add_op : xor_op], rst_op,no_op};
35          bins multi_cycle = {mul_op};
36
37          bins opn_rst[] = ([add_op:no_op] => rst_op);
38          bins rst_opn[] = (rst_op => [add_op:no_op]);
39
40          }
```

FIGURE 30-3. Multiple-Value Transitions

- **Line 37**—This line creates six bins, one for every operation followed by a reset.
- **Line 38**—This line creates six bins, one for a reset followed by every operation.

As we did before, we are not going to write a new test to make sure that our reset works in all cases. Instead, we'll let SystemVerilog automatically generate 100 transactions and run them through the self-checking test bench. Because we have a self-checking test bench, we know that if we've seen one of each of these combinations of operations, then they are all tested.

Here's the coverage after we run our 100 transactions:

Name	Coverage	Status	Auto_bin_max
bin opn_rst[no_op=>rst_op]	3		
bin opn_rst[rst_op=>rst_op]	5		
bin opn_rst[mul_op=>rst_op]	3		
bin opn_rst[xor_op=>rst_op]	2		
bin opn_rst[and_op=>rst_op]	3		
bin opn_rst[add_op=>rst_op]	6		
bin rst_opn[rst_op=>no_op]	4		
bin rst_opn[rst_op=>rst_op]	5		
bin rst_opn[rst_op=>mul_op]	3		
bin rst_opn[rst_op=>xor_op]	3		
bin rst_opn[rst_op=>and_op]	2		
bin rst_opn[rst_op=>add_op]	4		

FIGURE 30-4. Multiple-Value Coverage Results

Success! We've tested every operation before and after a reset.

30.2.1 Testing the Pipeline

Remember that we also needed to test that all single-cycle operations work properly before and after a pipelined multiply operation. This led to the following two coverpoints:

> *Have we run a single-cycle operation before a multicycle operation?*
> *Have we run a multicycle operation before a single-cycle operation?*

We need to capture these in our covergroup. We're now seeing how coverage-driven test plans work. We write them in English, but then we make executable coverage models so that the simulator can tell us whether we've met our coverage goals.

We keep using the coverpoint we created for the `op` variable. This is another key to using covergroups. The covergroups have `coverpoint` lines that relate to a variable we want to watch, and we keep adding bins to that coverpoint to get our complete coverage. You cannot have two `coverpoint` lines in the same covergroup that look at the same expression. All the bins associated with that expression need to be grouped together.

Here is our `op` coverpoint with new bins that check whether all the single operations were run before and after a multiply:

```
33      coverpoint op {
34         bins single_cycle[] = {[add_op : xor_op], rst_op,no_op};
35         bins multi_cycle = {mul_op};
36
37         bins opn_rst[] = ([add_op:no_op] => rst_op);
38         bins rst_opn[] = (rst_op => [add_op:no_op]);
39
40         bins sngl_mul[] = ([add_op:xor_op],no_op => mul_op);
41         bins mul_sngl[] = (mul_op => [add_op:xor_op], no_op);
42
43         }
```

FIGURE 30-5. Single-Cycle Operations Before and After Multiplies

- **Line 40**—This line uses a range to name the first three operations; then it skips over `mul_op` (not single-cycle) and `rst_op` (checked above) and adds in `no_op`. This line checks that all these single-cycle operations were followed by a multiply at least once.
- **Line 41**—This line makes sure that the multiply operation was followed by all the single-cycle operations at least once.

Once again, we don't change our test at all. We just run our 100 and see what happens:

Name	Coverage	Status	Auto_bin_max
bin sngl_mul[no_op=>mul_op]	3		
bin sngl_mul[xor_op=>mul_op]	2		
bin sngl_mul[and_op=>mul_op]	6		
bin sngl_mul[add_op=>mul_op]	2		
bin mul_sngl[mul_op=>no_op]	3		
bin mul_sngl[mul_op=>xor_op]	3		
bin mul_sngl[mul_op=>and_op]	3		
bin mul_sngl[mul_op=>add_op]	4		

FIGURE 30-6. Success! Tested Multiply Before and After Single-Cycle

Our little 100-transaction test bench is covering a lot of functionality!

30.3 Repetition Coverage

Sometimes the simplest things can slip through a test bench. I know an engineer who was burned by a bug in his chip that occurred whenever he ran the same operation twice in a row. In all his directed tests, it had never occurred to him to try the exact same operation twice in a row, and so this bug slipped through. We don't want that to happen to us, so let's add the following coverpoint to our test plan. We'll check all operations twice, and we'll run the pipelined multiply several times in a row:

Execute every operation twice in a row.
Run three to five multiplies in a row.

This is a simple coverpoint to describe, but it may not be such a simple one to capture. The most straightforward way to do this would be to create bins that look like this:

```
mul_op => mul_op
and_op => and_op
  and so on...
```

The problem is, this is a lot of typing, and it won't work if we change the operations in our design. We want to have something more general for describing repetition, and SystemVerilog fits the bill with two kinds of repetition operators:

- The `[* ]` consecutive repetition operator
- The `[= ]` and `[-> ]` nonconsecutive repetition operators

We'll use these to capture our repetition coverage requirements.

30.3.1 Consecutive Repetition

The consecutive repetition operator increments the bin if it sees the same data in a row a certain number of times. You supply the number. The bin looks like one of these syntaxes:

```
bins <name> = (<value list> [* n]);

bins <name> = (<value list> [* n:m]);
```

In this description, `n` and `m` are integers that tell us how many repetitions need to happen for us to increment the bin. We can use a single integer (`n`) or a range of integers (`n:m`)

Let's check for repetition in our test bench:

```
36
37        bins opn_rst[] = ([add_op:no_op] => rst_op);
38        bins rst_opn[] = (rst_op => [add_op:no_op]);
39
40        bins sngl_mul[] = ([add_op:xor_op],no_op => mul_op);
41        bins mul_sngl[] = (mul_op => [add_op:xor_op], no_op);
42
43        bins twoops[] = ([add_op:no_op] [* 2]);
44        bins manymult = (mul_op [* 3:5]);
45
46        }
```

FIGURE 30-7. Operation Repetition Bins

- **Line 43**—This line checks whether we ran all operations twice in a row.
- **Line 44**—This line checks whether we ran the multiply from three to five times in a row.

Here are the results of these bins run against 100 transactions:

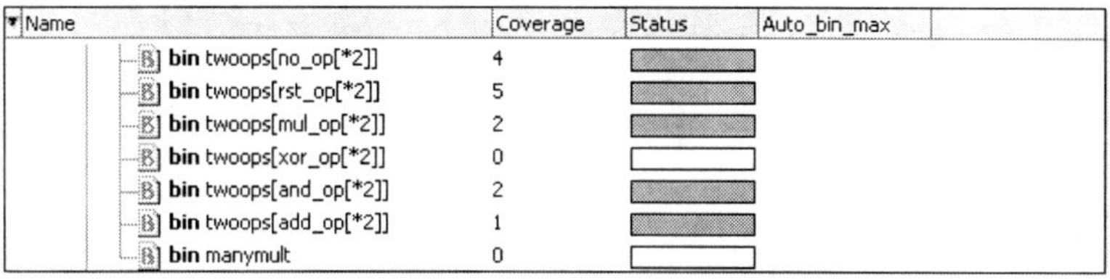

Name	Coverage	Status	Auto_bin_max
bin twoops[no_op[*2]]	4		
bin twoops[rst_op[*2]]	5		
bin twoops[mul_op[*2]]	2		
bin twoops[xor_op[*2]]	0		
bin twoops[and_op[*2]]	2		
bin twoops[add_op[*2]]	1		
bin manymult	0		

FIGURE 30-8. Results of Bin Repetition

We've finally found a limit to our 100-transaction test bench. It never ran two XOR operations in a row, and it never ran a multiply three to five times in a row.

We could solve this problem in one of three ways:

- We could add a test to the test bench that constrains the transactions to `mul_op` and `xor_op`. That would probably get us the repetition we want.
- We could try simulating with a different random seed, as we discussed in Section 20.4, "Seeding Randomization," on page 241.
- We could just run more transactions. When I changed the test bench to run 1000 transactions instead of 100, I got complete coverage.

The key was that functional coverage showed us a hole in our test bench, and gave us the opportunity to fix it.

30.3.2 Nonconsecutive Repetition

Where there is consecutive repetition we would expect to see nonconsecutive repetition and, sure enough, we have nonconsecutive repetition in SystemVerilog. Nonconsecutive repetition increments when it sees `n` number of values, regardless of whether there are other values in between.

For example, say we wanted to see the value 5 twice and we are analyzing two sequences:

```
1, 3, 5, 5, 3, 1
1, 5, 3, 5, 1, 3
```

Consecutive repetition would have counted only the top sequence, because we saw the value 5 twice in a row—but not in the bottom, because there was a 3 between each occurrence of a 5. Nonconsecutive repetition would have incremented on both.

But what purpose does nonconsecutive repetition serve? Does it add anything to have a special operator that tells us whether a value appeared `n` times somewhere in our test? That is nonconsecutive repetition, but we could have gathered that data by changing our coverage goal from 1 to `n`.

We don't use nonconsecutive repetition simply to count values. Instead, we use it as part of a longer sequence. Here is how we define nonconsecutive repetition, and then we'll see how to use it:

```
<list of values> [= n:m]

<list of values> [-> n:m]
```

Both of these operators deliver nonconsecutive repetition. If we put these operators into a bin by themselves, they both act the same way. They both increment if they see values from the `<list of values>` between `n` and `m` times. The difference between them appears when we use them in a bigger sequence.

Let's look at an example to explain the difference. Say we had a test plan coverpoint like this one:

Execute two or more `mul_op` *operations between a pair of* `rst_ops`.

We need to watch for a sequence where we see a reset, then one or two multiply operations, and then another reset. It is not clear from this description whether we need to see those operations immediately following each other or whether they can be spread out. If we want them immediately following each other, then we could capture this requirement very simply with the following:

```
rst_op => mul_op => mul_op => rst_op
```

That sequence is completely clear. Let's say that we brought that sequence to the requesting engineer who said, "No. No. They don't have to follow right after each other. All I need is to see a `rst_op`, then some operations including two `mul_ops`, then another `rst_op`."

Well, we're mildly chagrined, but we take solace in the fact that our engineer customer is not a very clear person—because he still hasn't told us what he wants. Clearly he wants the sequence to start with a `rst_op`, and clearly he doesn't only need to see `mul_ops` in the middle. Still, how does he want the sequence to end? Does he want that last `rst_op` to follow on the tail of the second `mul_op`, or can there be even more transactions between these last two? Annoyed, we decide to write it both ways, using the = and the -> operators:

```
45
46        bins rstmulrst    = (rst_op => mul_op [=  2] => rst_op);
47        bins rstmulrstim  = (rst_op => mul_op [-> 2] => rst_op);
48
49        }
```

FIGURE 30-9. Nonconsecutive Repetition with the TinyALU

- **Line 46**—The `[= ]` operator (called the *nonconsecutive* operator in the LRM) matches when it sees the `rst_op` after two `mul_ops`. It doesn't care how many intervening transitions happen after the second `mul_op`.
- **Line 47**—The `[-> ]` operator (called the *goto* operator in the LRM) matches only if a `rst_op` immediately follows the second `mul_op`.

These operators respond differently to the following sequences:

```
A) rst_op => ... => mul_op ... => mul_op ... => rst_op
B) rst_op => ... => mul_op ... => mul_op =>rst_op=>rst_op
```

The "..." can be any operation other than a `mul_op` or a `rst_op`.

The goto operator `[-> ]` matches sequence `B` only. It matches it because the `rst_op` immediately followed the `mul_op`.

The nonconsecutive operator `[= ]` matches both sequences.

We take these back to our requesting engineer, who then runs 300 transactions through the test bench, and we see the following coverage:

Name	Coverage	Status	Auto_bin_max
bin twoops[add_op[*2]]	1		
bin manymult	0		
bin rstmulrst	16		
bin rstmulrstim	2		

FIGURE 30-10. Results of Nonconsecutive Coverage

We can see that it is much easier to generate a sequence where the `rst_op` does not have to follow the `mul_op` immediately. We matched the *nonconsecutive* sequence 16 times in 100 transactions, but we matched the *goto* sequence only twice.

30.4 Summary

In this chapter we learned how to cover sequences of data in our test bench. We saw how to capture arbitrary sequences of values with the => operator. Then we saw how to capture consecutive sequences of values with the `[* ]`, `[= ]`, and `[-> ]` operators.

Finally, we saw that we can combine sequences with repetition operators to create long and complex sequences. We are doing all this so that we can create coverage-driven test plans that the simulator can track for us.

We can now cover a wide variety of values and sequences. Even so, we're still missing one important aspect of coverage: cross coverage. For example, we know that we've simulated with all zeros on the `A` leg of the TinyALU, and we know that we've simulated the multiply operation. But did we ever simulate the multiply operation with all zeros on one leg, or all ones on both legs?

Cross coverage is the last piece of functional coverage, and we'll cover it in our next chapter.

31 Cross Coverage

IN THIS CHAPTER:

- *Basic Cross Coverage*
- *Cross Coverage and the TinyALU*
- *Combining Cross-Coverage Bins*
- *The `intersect` Qualifier*
- *Implementing TinyALU Coverpoints*

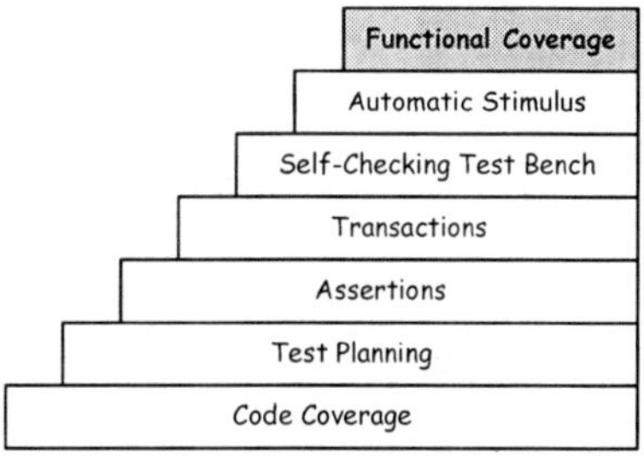

Let's say that you got tired of engineering and have moved on to new things. Specifically, you're now in charge of snacks for a summer camp. You have young campers who need to get a daily snack. You also have a pile of healthy snacks for the kids: apples, bananas, pears, and trail mix.

Your camp has instituted a "You have to at least try it" policy that requires that each kid try each snack once. By the end of the week, each kid needs to have had an apple, a banana, a pear, and some trail mix. Because there are five days in the camp week, kids can double up on favorites as long as they try each one.

Given a pencil, paper, and a clipboard (and perhaps a whistle for crowd control), you can easily do this job. You make a matrix with all the kids in rows and the four snacks in columns; when you give each kid a snack, you check off the appropriate cell in the matrix.

This is how cross coverage works. Cross coverage ensures that all the combinations of two variables have been covered. In this chapter, we'll learn how to create a *cross* in SystemVerilog covergroups.

31.1 Basic Cross Coverage

We are going to implement a model of our camper problem by using a SystemVerilog cross. A cross uses existing coverpoints to create a matrix of bins, and measures how many times we saw the data associated with each bin.

We use the keyword `cross` to create a cross, and we use it in the same place as the `coverpoint` keyword in a covergroup. The basic syntax looks like this:

```
<identifier> : cross <list_of_coverpoints>
```

There is an optional identifier and then the keyword `cross`. We follow that with a list of previously defined coverpoints. This is a subset of the complete declaration. (We'll discuss additional parts of the statement in later sections.) Let's use the SystemVerilog `cross` to model our camper problem.

We have three campers: Roger, Hui, and Marion. And we have four snacks: apples, pears, bananas, and trail mix. We need to check that all the campers get at least one of the snacks, so we're going to write a SystemVerilog model for this situation. There are four parts to the module:

- An enumerated type with the names of the campers.
- An enumerated type with the names of the snacks.
- A class that combines the kid and the snack. We'll use the class to randomize the snack/camper combinations.
- A covergroup to count the number of times we give a snack to each camper and the number of times we give each snack. Then we'll create a cross to see if everyone got one snack.

We randomize the snack object five times for each camper, to simulate giving the camper a snack each of the five days of camp.

Here's the code that implements our campers and snacks. It is a brand-new example of using objects, randomization, and covergroups, so let's go over it in detail:

```
16  typedef enum { roger, hui, marion} campers_t;
17  typedef enum {apple, pear, banana, trailMix} snacks_t;
18
19  class camperSnack;
20    randc campers_t camper;
21    rand  snacks_t snack;
22  endclass
23
24  module top;
25    campers_t camper;
26    snacks_t  snack;
27
28    covergroup clipboard;
29      coverpoint camper;
30      coverpoint snack;
31      matrix : cross camper, snack;
32    endgroup
33
34    initial begin
35      clipboard cb = new();
36      camperSnack snacktime = new();
37      repeat ( 5 * camper.num()) begin
38        assert(snacktime.randomize);
39        camper = snacktime.camper;
40        snack  = snacktime.snack;
41        cb.sample();
42      end
43    end
44  endmodule
```

FIGURE 31-1. Feed the Children

- **Line 16**—Represent the campers with an enumerated type.
- **Line 17**—Represent the snacks with a different enumerated type.
- **Lines 19–22**—Create a class that captures a camper/snack combination. The `camper` variable is a `randc`, so that we make sure all the campers get snacks.
- **Line 24**—We write a module to simulate this situation.
- **Lines 25–26**—We'll use these variables in the covergroup.
- **Line 28**—We need a covergroup to test that all the campers have had all the snacks. We call the covergroup `clipboard` to represent our clipboard with the matrix on it.
- **Line 29**—This `coverpoint` is connected to the `camper` variable. We let SystemVerilog create the automatic bins.
- **Line 30**—This `coverpoint` is connected to the `snack` variable.
- **Line 31**—We made a matrix on our clipboard and so we use that name for our cross. We are crossing the campers and the snacks.
- **Line 35**—We declare the variable `cb` to hold our covergroup and we create a new instance of `clipboard`.

- **Line 36**—Create a new `camperSnack` object, and put it into a variable called `snacktime`.
- **Line 37**—We are going to give each camper a snack. There are five days in a camp week, so we need to multiply the number of campers by five. The `randc` in the `camper` class makes sure that we distribute the snacks to the campers.
- **Line 38**—Randomize the `snacktime` object.
- **Line 39**—Read the `camper` out of the snacktime.
- **Line 40**—Read the `snack` out of snacktime.
- **Line 41**—Sample them with the `cp` covergroup.

A cross creates an *n*-dimensional array of all the bins in the crossed covergroups. In our case, the `camper` coverpoint creates three bins. The `snack` coverpoint creates four bins. This means that when we cross them we'll get a matrix with 12 entries (**3 x 4**). Let's run this simulation and see how we do.

First, let's look at the coverage on the camper coverpoint and the snack coverpoint to make sure that all the campers got five snacks and that we gave away all types of snacks:

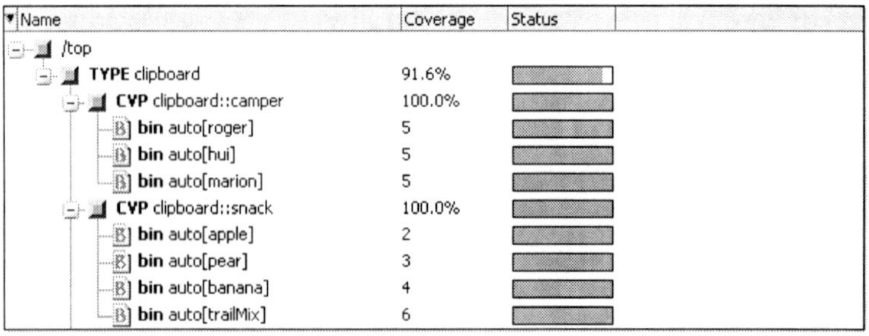

Name
Coverage
Status
/top
TYPE clipboard
91.6%
CVP clipboard::camper
100.0%
bin auto[roger]
5
bin auto[hui]
5
bin auto[marion]
5
CVP clipboard::snack
100.0%
bin auto[apple]
2
bin auto[pear]
3
bin auto[banana]
4
bin auto[trailMix]
6

FIGURE 31-2. Snacks and Campers Coverage

We can see that all the campers got five snacks. This was supposed to happen because of the `randc` variable, but it's nice to see that it really did. We also see that we gave away all the snacks at least once.

But there is trouble afoot. We gave out only two apples and there are three campers. Somebody didn't get an apple! Who is this unfortunate soul? Let's look at the cross coverage report to find out:

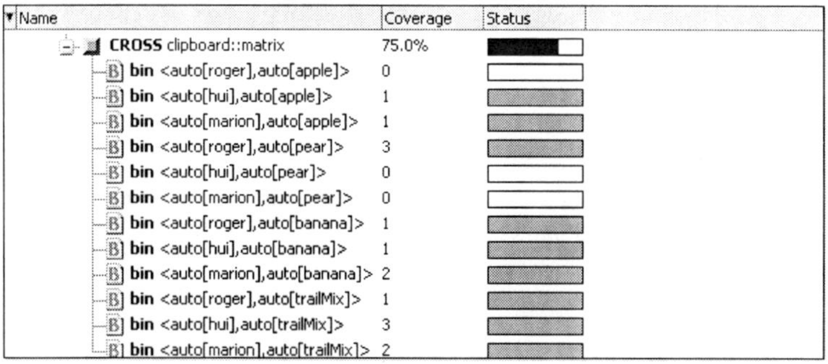

FIGURE 31-3. Camper/Snack Cross Coverage

The first cross bin shows us that poor Roger didn't get an apple. SystemVerilog has created an array of all the camper bins and all the snack bins, and then showed us every combination. We can see not only that Roger didn't get an apple, but also that Marion and Hui never got pears. In fact, it appears that Roger has stuffed himself with pears as he ate them three days out of five.

This example shows us the simplest form of cross coverage. A cross gives us a matrix of bins, and we can see whether we've touched all the combinations of bins.

This example also shows how quickly cross coverage can grow. We had only two variables in the cross and it generated twelve bins, showing how one can easily create cross-coverage models that outgrow the memory in most computers. For example, crossing all the addresses in a memory against read and write operations would be a very bad idea.

31.2 Cross Coverage and the TinyALU

Now that we have cross coverage, we can add three more coverpoints to our -TinyALU coverage-driven test plan:

Execute every operation with all ones on one of the inputs.

Execute every operation with all zeros on one of the inputs.

Execute the multiply with all ones on both inputs.

To do this we need to create three coverpoints: one for the `A` leg of the TinyALU, one for the `B` leg, and one for the operation. To avoid confusing this set of operation bins with the ones we used in previous chapters, we are going to create a new covergroup in the `alu_tester` module:

```
63 covergroup zeros_and_ones;
64
65    all_ops : coverpoint op {
66       ignore_bins null_ops = {rst_op, no_op};}
67
68
69    a_leg: coverpoint A {
70       bins zeros = {'h00};
71       bins others= {['h01:'hFE]};
72       bins ones  = {'hFF};}
73
74    b_leg: coverpoint B {
75       bins zeros = {'h00};
76       bins others= {['h01:'hFE]};
77       bins ones  = {'hFF};}
78
79    basic : cross a_leg, b_leg, all_ops;
```

FIGURE 31-4. Covergroup to Check for Zeros and Ones

- **Line 63**—Defining a covergroup called `zeros_and_ones`.
- **Lines 60–61**—Create a coverpoint for the `op` variable. We don't care about `no_op` and `rst_op` operations so we ignore them with the `ignore_bins` statement. The coverpoint automatically creates bins for the rest of the values.
- **Line 63–71**—Create coverpoints for `A` and `B` with identical bins. We have one bin for all ones, a second for all zeros, and a third for everything else.
- **Line 73**—The meat of the covergroup. Cross `A`, `B`, and `op` to create a three-dimensional array of bins.

When we run 50 transactions through our test bench (with a constraint that gives us many `'h00` and `'hFF` values) to get the following coverage:

Name	Coverage	Status
CROSS zeros_and_ones::basic	66.6%	
bin <zeros,zeros,auto[add_op]>	1	
bin <zeros,zeros,auto[and_op]>	0	
bin <zeros,zeros,auto[xor_op]>	1	
bin <zeros,zeros,auto[mul_op]>	2	
bin <others,zeros,auto[add_op]>	2	
bin <others,zeros,auto[and_op]>	0	
bin <others,zeros,auto[xor_op]>	0	
bin <others,zeros,auto[mul_op]>	1	
bin <ones,zeros,auto[add_op]>	3	
bin <ones,zeros,auto[and_op]>	1	
bin <ones,zeros,auto[xor_op]>	1	
bin <ones,zeros,auto[mul_op]>	0	
bin <zeros,others,auto[add_op]>	0	
bin <zeros,others,auto[and_op]>	0	
bin <zeros,others,auto[xor_op]>	0	
bin <zeros,others,auto[mul_op]>	1	
bin <others,others,auto[add_op]>	1	
bin <others,others,auto[and_op]>	1	
bin <others,others,auto[xor_op]>	1	
bin <others,others,auto[mul_op]>	1	
bin <ones,others,auto[add_op]>	2	
bin <ones,others,auto[and_op]>	0	
bin <ones,others,auto[xor_op]>	0	
bin <ones,others,auto[mul_op]>	1	
bin <zeros,ones,auto[add_op]>	3	
bin <zeros,ones,auto[and_op]>	1	
bin <zeros,ones,auto[xor_op]>	0	
bin <zeros,ones,auto[mul_op]>	1	
bin <others,ones,auto[add_op]>	0	
bin <others,ones,auto[and_op]>	0	
bin <others,ones,auto[xor_op]>	2	
bin <others,ones,auto[mul_op]>	4	
bin <ones,ones,auto[add_op]>	1	
bin <ones,ones,auto[and_op]>	1	
bin <ones,ones,auto[xor_op]>	1	
bin <ones,ones,auto[mul_op]>	3	

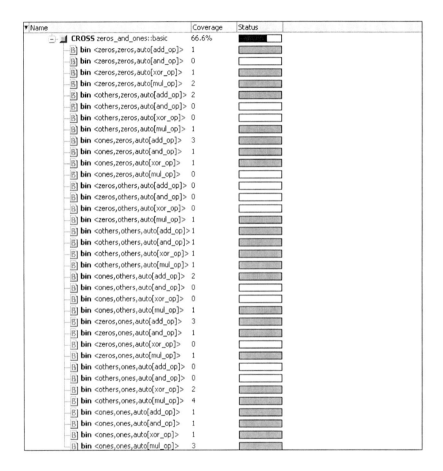

FIGURE 31-5. Cross-Coverage Results for Ones and Zeros

According to this we've barely tested our design. This is misleading. If we look more closely at the cross, we see that we've covered all the operations, and we've had plenty of situations where the A leg or the B leg had all ones or all zeros. Why does our coverage look so incomplete?

The problem is that we have too many bins. The cross blindly created 36 bins **(3x3x4)**. Then SystemVerilog measures how much we've covered those 36 bins. We needed only nine bins to fulfill the test plan. The rest of these bins are extra and we should get rid of them.

31.3 Combining Cross-Coverage Bins

In Figure 31-5 on page 367 we were trying to check the following coverage requirements:

Execute every operation with all ones on one of the inputs.

Execute every operation with all zeros on one of the inputs.

Execute the multiply with all ones on both inputs.

We create a basic cross of three coverpoints and ran a simulation with 50 transactions. When we first looked at the results, we saw that we had not achieved 100% coverage on our covergroup; but when we looked more closely, we saw that we had achieved our coverage requirements.

What's wrong? The problem is that our covergroup doesn't accurately reflect our coverage requirements. It is implementing the following:

Simulate every operation with every combination of all zeros, all ones, and other numbers.

If we look at our covergroup results in terms of this rule, we see that we did not implement it. For example, we did not test the AND operation with zeros on the `A` and `B` legs.

Making cross coverage more precise is a lot like sculpting. We start with the raw material—all the combinations that come from crossing our coverpoints. This creates a large number of bins. We then create the coverage we want by combining these bins.

Let's examine this process by looking at our coverage matrix graphically. The `basic` cross-coverage matrix looks like this for the TinyALU:

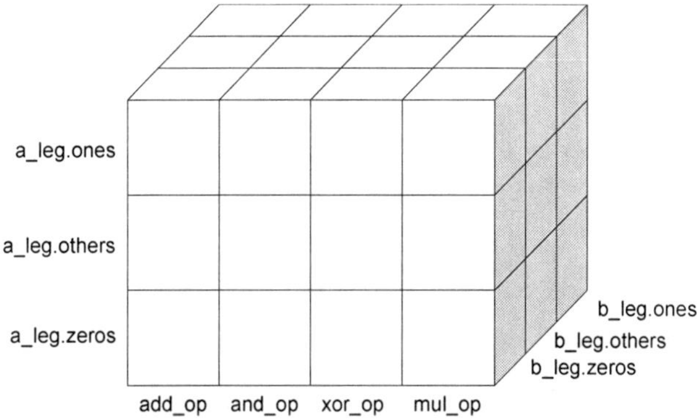


```
covergroup zeros_and_ones;

   all_ops : coverpoint op {
      ignore_bins null_ops = {rst_op, no_op};}


   a_leg: coverpoint A {
      bins zeros = {'h00};
      bins others= {['h01:'hFE]};
      bins ones  = {'hFF};}

   b_leg: coverpoint B {
      bins zeros = {'h00};
      bins others= {['h01:'hFE]};
      bins ones  = {'hFF};}

   basic : cross a_leg, b_leg, all_ops;
```

FIGURE 31-6. Default Cross for the TinyALU

There are 36 bins in this matrix. SystemVerilog defines each bin in terms of the three bins that intersect to form it. For example, in Figure 31-5 we see that the first bin in the cross is defined by `a_leg.zeros`, `b_leg.zeros`, and `op.add_op`. We use these bin names to tell SystemVerilog how to sculpt our coverage matrix.

The bins in our basic matrix were all generated automatically. It is time to replace them with more useful bins. Eventually, we want to create bins that implement our coverage model exactly.

Our first bin catches all the cases where the TinyALU's A leg is all zeros. The picture shows us that there are 12 bins in which this is the case. There are three different bins for the B leg and four bins for the operations.

We define bins explicitly with the binsof keyword. The syntax for the binsof definition looks like this:

```
binsof (<coverpoint_bin>)
```

The binsof selector tells SystemVerilog to select all the cells in the matrix that are touched by a bin. We want all the "bins of" the a_leg.zero bin, for example.

Here's an example of using binsof. In this case we've combined all the cross bins that use the a_leg.zeros bin into one big bin.

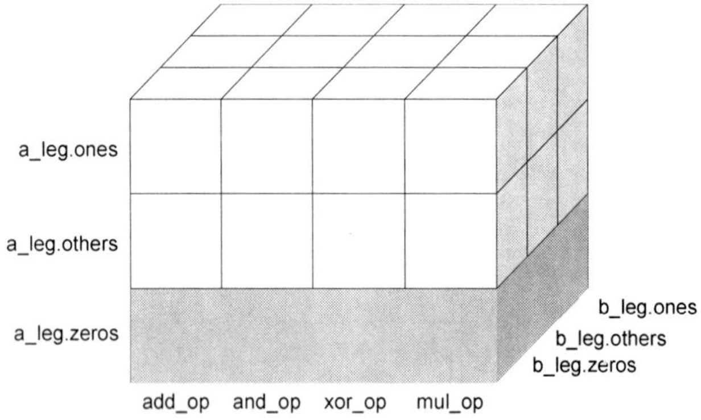


```
    justa0: cross a_leg, b_leg, all_ops {
       bins  a0bin = binsof(a_leg.zeros);}
```

FIGURE 31-7. Creating a Large Cross Bin

We use the bins operator to create bins within the cross, and then use the binsof operator to combine the a_leg.zero bins.

When we run 50 transactions through this covergroup, we see the following:

Name	Coverage	Status
CROSS zeros_and_ones::justa0	72.0%	
bin <others,zeros,auto[add_op]>	2	
bin <others,zeros,auto[and_op]>	0	
bin <others,zeros,auto[xor_op]>	0	
bin <others,zeros,auto[mul_op]>	1	
bin <ones,zeros,auto[add_op]>	3	
bin <ones,zeros,auto[and_op]>	1	
bin <ones,zeros,auto[xor_op]>	1	
bin <ones,zeros,auto[mul_op]>	0	
bin <others,others,auto[add_op]>	1	
bin <others,others,auto[and_op]>	1	
bin <others,others,auto[xor_op]>	1	
bin <others,others,auto[mul_op]>	1	
bin <ones,others,auto[add_op]>	2	
bin <ones,others,auto[and_op]>	0	
bin <ones,others,auto[xor_op]>	0	
bin <ones,others,auto[mul_op]>	1	
bin <others,ones,auto[add_op]>	0	
bin <others,ones,auto[and_op]>	0	
bin <others,ones,auto[xor_op]>	2	
bin <others,ones,auto[mul_op]>	4	
bin <ones,ones,auto[add_op]>	1	
bin <ones,ones,auto[and_op]>	1	
bin <ones,ones,auto[xor_op]>	1	
bin <ones,ones,auto[mul_op]>	3	
bin a0bin	10	

FIGURE 31-8. Cross Coverage with all `a_leg.zero` Bins Combined

Our coverage has improved! When we had 36 bins our coverage was 67%, and now it's 72%. The reason it's better is that we only need to touch one of the 12 `a_leg.zero` bins to get credit for all of them. In the default case, we could only get 100% coverage if all 36 bins have been touched. Now we can get 100% coverage if the large `a0bin` is touched and the remaining 24 automatic bins are touched.

31.3.1 Combining `binsof` Operations

The `binsof` operator allows us to create big cross bins from automatically generated bins. In our previous section, we combined all the bins together where the `a_leg` was all zeros.

Our coverage plan calls for us to test the cases where the TinyALU's A-leg or B-leg are all zeros. This calls upon us to combine bins using the binsof operator along with logical operations.

We can combine binsof operators with the AND (&&) and OR (||) operators. The syntax looks like these:

```
bins <bin> = binsof(<somebins>) && binsof(<otherbins>);

bins <bin> = binsof(<somebins>) || binsof(<otherbins>);
```

The first of these examples creates a smaller bin that counts only the intersection of the bins defined by the first binsof operation and the second binsof operation.

The second example creates a larger bin by combining the bins designated by the first and second binsof operations.

Let's use the || logical operator to get closer to the coverpoints in our plan. The plan says that we need to have zeros or ones on either the TinyALU's A-leg or B-leg. We have cross coverpoints that handle zeros and ones on both legs, so let's combine them.

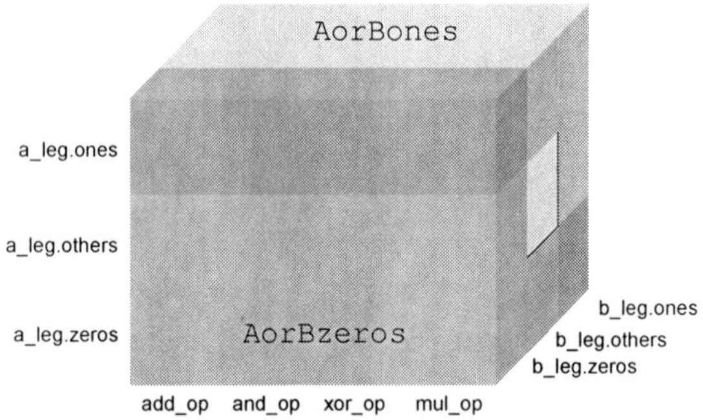


```
zeros_or_ones: cross a_leg, b_leg, all_ops {
   bins  AorBzero = binsof(a_leg.zeros) || binsof(b_leg.zeros);
   bins  AorBones = binsof(a_leg.ones ) || binsof(b_leg.ones );}
```

FIGURE 31-9. Creating Bins to Catch All Ones and All Zeros

When we create bins like this, we reduce the number of bins significantly. Now we'll see how many times we had all ones or all zeros on the TinyALU's `A`-leg or `B`-leg. Here's the simulation result:

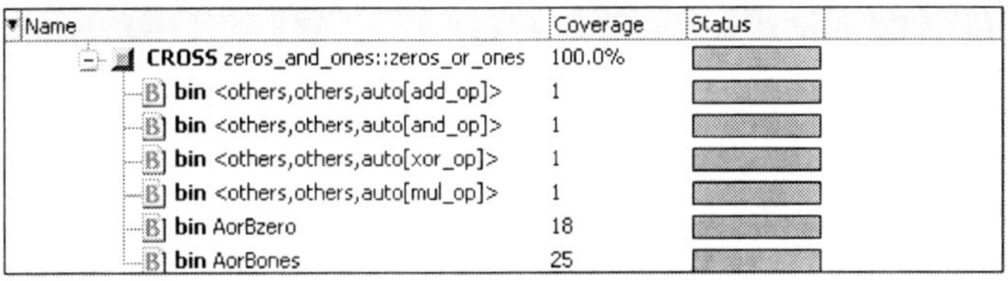

Name	Coverage	Status
CROSS zeros_and_ones::zeros_or_ones	100.0%	
bin <others,others,auto[add_op]>	1	
bin <others,others,auto[and_op]>	1	
bin <others,others,auto[xor_op]>	1	
bin <others,others,auto[mul_op]>	1	
bin AorBzero	18	
bin AorBones	25	

FIGURE 31-10. Finding Zeros and Ones on `A` and `B`

We now see that we've got only six bins left. There is one bin that captures cases where there are zeros on `A` or `B`, and another bin that catches the case where there are ones on `A` or `B`.

The bins in Figure 31-9 give us the square donut in Figure 31-9 on page 372. The four unused bins in the middle are the `<others, others>` bins. Even though we have 100% coverage, it is imprecise to leave these bins in our covergroup when we don't care about them in our report. So let's remove them.

31.3.2 Ignore Bins

Our goal is to create a coverage model that meets our needs exactly, and so we want to ignore the four `<others, others>` bins in Figure 31-9. We do that with the `ignore_bins` statement.

The `ignore_bins` statement works just like the `bins` statement, but SystemVerilog removes the bins in the statement from the coverage model.

Here is how we use `ignore_bins` in the TinyALU:

```
81    zeros_or_ones: cross a_leg, b_leg, all_ops {
82       bins  AorBzero = binsof(a_leg.zeros)  || binsof(b_leg.zeros);
83       bins  AorBones = binsof(a_leg.ones )  || binsof(b_leg.ones );
84       ignore_bins the_others
85                      = binsof(a_leg.others) && binsof(b_leg.others);}
```

FIGURE 31-11. Removing Bins by Using `ignore_bins`

- **Lines 84–85**—We use the `&&` operator along with the `binsof` selectors to say that we ignores all bins made from both `a_leg.others` and `b_leg.others`.

The `ignore_bins` operator allows us to create a very precise coverage model. Now when we run the simulation we see the following:

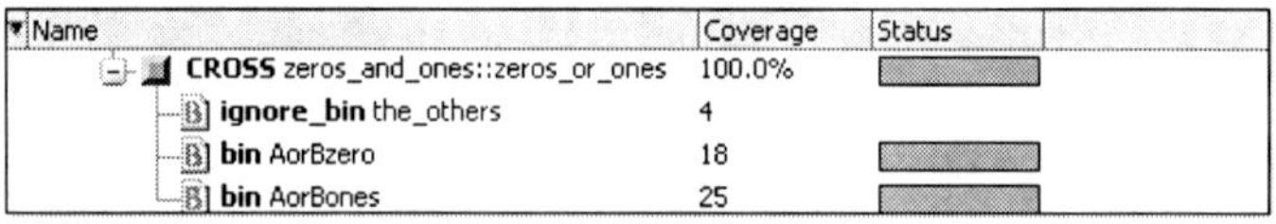

FIGURE 31-12. Coverage with Ignored Bins

The bin `the_others` is excluded from our coverage model. We are almost there when it comes to creating that coverage model. We can now see how many times we've had `'h00` or `'hFF` on the `A` or `B` ports. However, our coverage model asked for more than this. We need to be able to see whether we've checked every operation. We have bins to capture the operations we run, but SystemVerilog automatically created these (as opposed to our defining them.) This means that we'll need an additional tool to go along with the `binsof` selector.

31.4 The `intersect` Qualifier

In Figure 31-7 on page 370 we used the `binsof` operator to aggregate coverage for all bins touched by `a_leg.zeros`. We were able to do this because we had explicitly defined the zeros bin in our `a_leg` coverpoint definition. However, we can't use the same method for the `ops` coverpoint, because its bins were created automatically. In this case we'll have to choose bins based on the values they cover.

SystemVerilog allows you to create a bin based on values with the `intersect` modifier to the `binsof` keyword. You add `intersect` to a `binsof` selector like this:

```
binsof (<bin>) intersect (<value_list>)
```

The <value_list> can be a single value, a list of values, or a range of values stored in square brackets.

Here is an example of using the `intersect` keyword to gather data about the `add_op` bin:

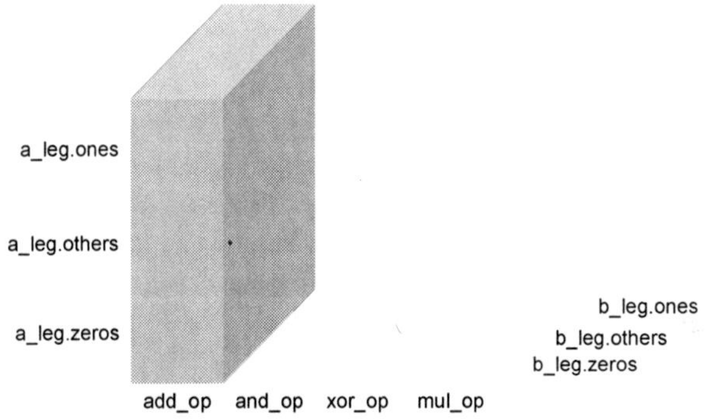


```
87    add_ops: cross a_leg, b_leg, all_ops {
88       bins add_bin = binsof(all_ops) intersect {add_op};
89       ignore_bins x= ! binsof(all_ops) intersect {add_op};
90    }
```

FIGURE 31-13. Using Intersect to Create a Bin

- **Line 88**—Use the `intersect` keyword to select all automatically generated `add_op` bins that *do* touch the value `add_op`.
- **Line 89**– Use the `!` qualifier to select all automatically generated `all_op` bins that *don't* touch the value `add_op`.

This example demonstrates two ways to use the `intersect` modifier for `binsof`. The `add_bins` bin uses `intersect` to choose the values that match the list. The `ignore_bin` uses `intersect` to choose the values that don't match the list. The exclamation point `(!)` inverts the bins chosen by `intersect`.

Important: The `(!)` *works only with the values of* `intersect`. *It has no effect on a* `binsof` *statement that has no* `intersect`.

We now have enough tools to implement the coverpoints from our test plan.

31.5 Implementing the TinyALU Coverpoints

With enough tools to implement the TinyALU coverpoints, can capture the following three coverpoints from our verification plan:

Execute every operation with all ones on one of the inputs.

Execute every operation with all zeros on one of the inputs.

Execute the multiply with all ones on both inputs.

These coverpoints suggest that we'll need a covergroup with two bins per operation. We'll have one bin that counts the number of times we see the operation with `'hFF` on one of its inputs, and another bin that counts the number of times we see the operation with `'h00` on one of its input. The same transaction would be counted twice if we see `'hFF` on one leg and `'h00` on the other leg.

We'll also create a special bin for the multiply operation with `'hFF` on both legs. This is a corner case, and it creates the biggest number possible from the TinyALU.

We'll use the `binsof` selector we discussed in the previous sections to choose bins with `'h00` or `'hFF`. Then we'll use the `intersect` selector to combine these bins with specific operations.

The complete covergroup looks like this:

```
53 covergroup zeros_or_ones_on_ops;
54
55    all_ops : coverpoint op {
56       ignore_bins null_ops = {rst_op, no_op};}
57
58    a_leg: coverpoint A {
59       bins zeros = {'h00};
60       bins others= {['h01:'hFE]};
61       bins ones  = {'hFF};}
62
63    b_leg: coverpoint B {
64       bins zeros = {'h00};
65       bins others= {['h01:'hFE]};
66       bins ones  = {'hFF};}
67
68    op_00_FF:  cross a_leg, b_leg, all_ops {
69       bins add_00 = binsof (all_ops) intersect {add_op} &&
70                   (binsof (a_leg.zeros) || binsof (b_leg.zeros));
71
72       bins add_FF = binsof (all_ops) intersect {add_op} &&
73                      (binsof (a_leg.ones) || binsof (b_leg.ones));
74
75       bins and_00 = binsof (all_ops) intersect {and_op} &&
76                   (binsof (a_leg.zeros) || binsof (b_leg.zeros));
77
78       bins and_FF = binsof (all_ops) intersect {and_op} &&
79                      (binsof (a_leg.ones) || binsof (b_leg.ones));
80
81       bins xor_00 = binsof (all_ops) intersect {xor_op} &&
82                   (binsof (a_leg.zeros) || binsof (b_leg.zeros));
83
84       bins xor_FF = binsof (all_ops) intersect {xor_op} &&
85                      (binsof (a_leg.ones) || binsof (b_leg.ones));
86
87       bins mul_00 = binsof (all_ops) intersect {mul_op} &&
88                   (binsof (a_leg.zeros) || binsof (b_leg.zeros));
89
90       bins mul_FF = binsof (all_ops) intersect {mul_op} &&
91                      (binsof (a_leg.ones) || binsof (b_leg.ones));
92
93       bins mul_max = binsof (all_ops) intersect {mul_op} &&
94                      (binsof (a_leg.ones) && binsof (b_leg.ones));
95
96       ignore_bins others_only =
97                    binsof(a_leg.others) && binsof(b_leg.others);
98       }
99 endgroup
100
```

FIGURE 31-14. TinyALU Zero and One Covergroups

- **Line 53**—Defining a covergroup called `zero_or_ones_on_ops`.
- **Line 55**—Create bins for all the operations except the `rst_op` and `no_op`.
- **Lines 58–61**—Creates three bins for the `A` input to the TinyALU. One looks for `'h00`, one looks for `'hFF`, and the last one looks for other values.
- **Lines 63–66**—Do the same thing for the B leg.
- **Line 68**—Create a cross for the `a_leg`, `b_leg`, and `all_ops` coverpoints. This creates 36 bins that we need to carve up to get our coverage.

- **Lines 69–70**—We want to catch the case where we see `'h00` on either leg and we are doing an ADD operation. Implement this by ORing together the bins that contain `'h00` on either leg, then ANDing the result with a bin that selects only the `add_op`.
- **Lines 72–73**—Count the times we see `'hFF` on either leg with an `add_op` operation.
- **Lines 78–91**—Repeat this pattern for all the operations. Creating two bins for each operation: one that looks for `'h00` and one that looks for `'hFF`.
- **Lines 93–94**—Create a special bin that looks for the multiply operation and `'hFF` on both legs.
- **Lines 96–97**—Ignore bins where we don't have `'h00` or `'hFF` on either leg.

When we run our 50 transactions through the test bench, we see the following results:

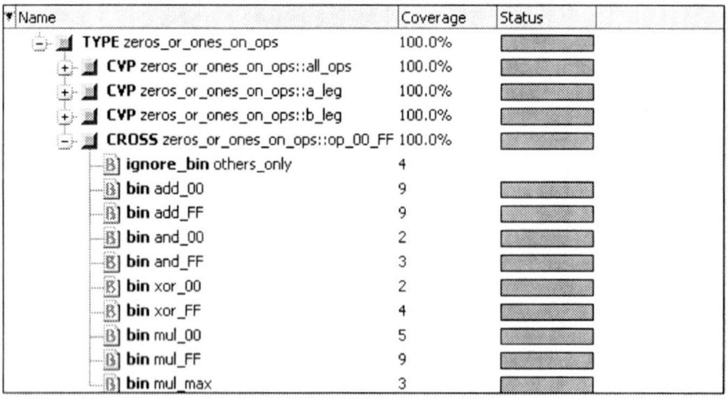

FIGURE 31-15. Coverage Results for the Zeros and Ones Requirement

We now see that with only 50 transactions and the right constraints, we were able to test our all-zeros and all-ones requirement on every operation.

We also tested the maximum number that could come out of the TinyALU three times.

31.6 Summary

In this chapter we learned how to create covergroup crosses. These are combinations of coverpoints that show whether we've covered every combination of values in one coverpoint's bins with all the bins in another coverpoint. We saw that we could use crosses to complete our description of a cover model for a coverage-based test plan.

Then we saw that we could control coverage further by combining the bins in a cross to make larger bins, or to exclude bins. We examined the `binsof` keyword and its counterpart, the `intercept` keyword.

We can combine these operators with the `&&`, `||` and `!` operators to create logical expressions that further combine and bind bins out of the basic bins in a cross.

Cross coverage gives us the ability to ensure that we've tested all the interesting combinations of transactions in a design. It is the final piece of functional coverage.

We've reached the top of the staircase.

32 Take the First Step

In This Chapter:

- *Simulate Today!*
- *Join the Community*

Predicting the future is easy, as long as you pick the right question.

It's hard to predict the future if there are too many forces at work. No futurists predicted that video phones would be dead, but that instead we'd have websites devoted to up uploading goofy movies of ourselves. There were too many conflicting forces to see it coming.

Then again, it would have been easy to predict that prices would be higher today than they were 20 years ago. The forces in this case are simple and unstoppable.

The future of FPGA design is easy to predict. There is only one rule:

Chips are getting bigger and chips are getting faster.

This simple rule guarantees that we'll soon be unable to do interesting FPGA designs without simulation. The chips will be too big and too complex to debug in the lab. This leads to a simple prediction about the near future:

Successful FPGA designers will simulate their FPGAs.

It's that simple. While FPGA simulation may seem like a novelty or a luxury today, it will be a basic staple of design in the future.

Years ago, when I was a college hire, I knew an engineering manager who opposed the latest EDA technology: computer-aided schematic capture. He believed that logic diagrams should be drawn on paper, and he could point to one board that was late because the engineer had missed connecting a signal by a single pixel. His favorite scheduling quote was, "We can do that in three weeks, or in six weeks if you make us use CAD."

He was let go in the first 1980's layoff.

FPGA simulation today is like computer-aided design was twenty years ago. It is new to some engineering teams, but ten years from now it will be standard operating procedure. No serious engineer will try to debug an FPGA without simulating it, and engineering managers who oppose simulation will be the subject of stories like the one above.

William Wulf, president of the National Academy of Engineers, said the following in a *Christian Science Monitor* article about a round of layoffs from Nortel:[1]

> *"The half-life of engineering knowledge, the time it takes for something to become obsolete, is from seven to 2–1/2 years. Lifelong learning is critical in this profession."*

This means that, at best, half your skills will be obsolete in seven years. Given the increases in complexity in FPGA designs, it's pretty clear that lab debug skills will be among the obsolete ones. The key to success as an engineer is to pick up the new skills before they become critical.

32.1 Simulate Today!

As we learned in college, the key to learning new engineering skills is to use new engineering skills. We engineers are, at heart, practical people. Theory goes only so far with us, and eventually we want to get down to actually using what we learned. The key to learning how to simulate your FPGAs is, well, to simulate your FPGAs.

I wrote this book in a step-by-step fashion so that you could use it no matter where you are on the simulation curve. If you're just simulating a little today, then you have an easy first step: just turn on code coverage.

1. "A short circuit for US engineering careers," *Christian Science Monitor*, December 26, 2002.

The key is just to start using these techniques. And when you do start using these techniques, remember this: you will get it wrong the first time. We all do.

32.1.1 Lousy First Drafts

I have to be honest, you're holding the fourth draft of this book. I literally wrote this book three times. The first draft was terrible. Absolutely terrible. I never even showed it to anyone.

The second draft was better. It was so good that I thought it might be the final draft. Then I showed it to people and I learned that it was also terrible.

The third draft was almost on the money. It was accessible and readable, but it was still too complex.

You're holding the fourth draft.[2]

The key to this story is that we always get things wrong when we do them the first time. Our first test plans will be off base. They will need to be rewritten completely. Our first assertions will not actually catch all the bugs we intended. Or they will catch bugs that don't exist. Our first transaction-level test bench just won't work. None of mine did.

We learn new skills by embracing our crappy first drafts. They help us internalize the skills so that the next time we use them, things go more smoothly. We learn by making mistakes. It's always been that way.

This is not to say our first efforts will be useless. While our first test bench won't be as well-designed as our third one, it will still be better than not learning the skill at all. The first time we turn on code coverage we'll be appalled, but we'll learn how to improve our coverage for the next project.

32.2 Join the Community

Recently, I installed Ubuntu Linux on my home computer. Naturally, I ran into problems. That's when I discovered the forums devoted to helping people get

2. Which may still be terrible—but darn it, I'm done!

started on Ubuntu. Every question I asked was answered immediately by an expert somewhere in the world.

Take advantage of a community like this when you learn about FPGA simulation. The website www.fpgasimulation.com is a good place to start. You can go there to join a forum dedicated to FPGA simulation. You can also find example designs, tips, and articles.

Once you've had some questions answered and learned some new skills, then you can be a resource for the next batch of learners. There's nothing like answering questions to help you learn what you thought you already knew. Eventually, you may want to provide code that you've written to help others verify their FPGAs.

Being seen on the Internet as an expert also makes for great job security!

32.3 Summary

It's clear that FPGA simulation is here to stay, and it will only become more important as time goes by. I hope you'll use the skills introduced in this book. The book contains everything you need to simulate your FPGAs and get them out the door successfully. You can absolutely learn to do this. It's easy, and you're smart.

Take the first step.

Index

U

V

W

Y

Printed in the United States
140539LV00003B/2/P

9 780974 164908